D1255561

FISH
Antifreeze
PROTEINS

Molecular Aspects of Fish and Marine Biology – Vol.1

FISH Antifreeze PROTEINS

editors

Kathryn Vanya Ewart

National Research Council Institute for Marine Biosciences, Canada

Choy Leong Hew

National University of Singapore, Singapore

World Scientific

New Jersey • London • Singapore • Hong Kong

Published by

World Scientific Publishing Co. Pte. Ltd.
P O Box 128, Farrer Road, Singapore 912805
USA office: Suite 1B, 1060 Main Street, River Edge, NJ 07661
UK office: 57 Shelton Street, Covent Garden, London WC2H 9HE

Library of Congress Cataloging-in-Publication Data
Fish antifreeze proteins / editors, Kathryn Vanya Ewart, Choy Leong Hew.
p. cm. -- (Molecular aspects of fish and marine biology ; v. 1)
Includes bibliographical references and index.
ISBN 9810248997 (alk. paper)
1. Antifreeze proteins. 2. Fishes--Composition. I. Ewart, Kathryn Vanya. II. Hew, Choy L. (Choy Leong), 1942– III. Series.

QP552.A56 F55 2002
572'.617--dc21 2002071344

British Library Cataloguing-in-Publication Data
A catalogue record for this book is available from the British Library.

This book is printed on acid-free paper.

Printed in Singapore by Uto-Print

Preface to the Series on *Molecular Aspects of Fish and Marine Biology*

This monograph is the first in a series on the molecular biology of fishes. Fish are becoming an increasingly significant focus in molecular biology and biotechnology. Several fish species have also emerged as valuable model systems for medical and general life sciences research. Prominent examples include the zebrafish in developmental biology and the puffer fish in genomics research. Fish are even more important commercially, mainly as a food source but also as ornamental or pet species. Research on fish using current biochemical, molecular and genomic approaches is essential in order to develop optimal health care and productivity for both cultured and wild species. The goal of this series is to review current developments in key areas of fish molecular biology that impact on our understanding of fish biology and have relevance in areas ranging from fisheries and aquaculture to medical research.

Preface to the Monograph on *Fish Antifreeze Proteins*

The inaugural monograph in this series is on the antifreeze proteins and glycoproteins (AF(G)Ps) of marine fishes. The AF(G)P is the key example of a discovery first made in fish that has had a wide impact on various areas of life sciences and biotechnology. The AF(G)Ps were initially considered as a remote biological curiosity in Antarctic teleosts. With further research, the occurrence of AF(G)Ps was quickly found to extend far beyond the Antarctic into the North temperate and Arctic oceans and the diversity of species known to produce them grew as well. These varied proteins and their genes have now emerged as a series of distinct and valuable models in various areas of life sciences research. Most exciting, perhaps, is that the AF(G)Ps are now leading to promising new biotechnology applications. The goal of this monograph is to provide a complete and concise reference representing current understanding of fish AF(G)Ps.

At the core of the study of fish AF(G)Ps lies a compelling mystery. The precise mechanism of ice binding by these proteins is something that, for 30 years, has remained elusive. Even with the more recent discovery of AF(G)Ps from insects and plants, the structure-function relationship is not yet clearly described. But in studies ranging from atomic protein structure to fish population dynamics, studies involving AF(G)Ps have led to new insights, innovative tools and original ideas. In the quest for an understanding of the molecular action of AF(G)Ps and of their biological roles and regulation, researchers have sometimes uncovered what they sought and sometimes not. Frequently, the studies seeking to understand various aspects of the AF(G)Ps led to valuable new developments in completely different research areas.

The discovery of AF(G)Ps began, like many others, with a perceptive question about fish adaptation. This led to a remarkable odyssey in the Antarctic that is recounted here in the chapter by Robert Feeney and David Osuga. Groundbreaking work by Robert Feeney, Arthur DeVries, and many other researchers on the freezing point depression of Antarctic fishes led to the discovery of AFGPs. This research set the stage for the many developments that followed. Study of the recent evolution of the Antarctic AFGPs revealed the *de novo* synthesis of a gene for AFGP from an internal exon-intron boundary within a trypsinogen gene. This represents a profound paradigm shift in our understanding of gene evolution because the new AFGP emerged by repetition and expansion of a non-coding element. It is made even more remarkable by the isolation of an evolutionary intermediate, a trypsinogen gene that contains the functional AFGP. Thus, it has provided a rare glimpse of molecular evolution in action. The AFPs have also represented unusual structures of interest in protein chemistry and functional studies. The crystal structure of a type I AFP was found to be the longest single alpha helix characterized at this level. The crystal structure of the ocean pout AFP, in contrast, was shown to be a previously unknown protein fold with one very flat surface upon which the ice ligands were identified. Regulation of the winter flounder AFP expression by temperature through the negative action of growth hormone has been shown to involve a common cellular signal transduction pathway but with some interesting and unique components. The role of a putative helicase in the regulation of winter flounder AFP gene expression is a surprising addition to the knowledge of seasonal gene regulation. Winter flounder AFPs also exist in an amazing variety including some forms predominantly produced in liver and secreted into the blood and others that are more abundant in skin, appearing to be intracellular. The molecular and biological functions of the skin AFPs are described and discussed in this volume. Another facet of AF(G)P research is their role in fish survival in cold oceans. A physiological, oceanographic and population-level investigation of AF(G)P in species of the North Atlantic ocean is included here as well. Finally, a brief synthesis of our understanding of AF(G)Ps from the whole animal to the molecular context with a review of their measurement and analysis is provided aiming to draw the vast areas of investigation that comprise fish AF(G)P research into a common theme.

The AF(G)Ps and their genes are finding applications in biotechnology. A promising development in AF(G)Ps research is their use in the production of transgenic fish. Although freeze resistance has not yet emerged as a phenotype from this work, the genetics are in place with stable heritable transgenes for winter flounder AFP in the genome of Atlantic salmon. The most successful aspect of this work is the production of growth hormone transgenic fish using the ocean pout antifreeze protein gene promoter to drive expression. This successful transgene technology for fish was first developed in pioneering experiments with antifreeze protein genes. It has marked the beginning of a new approach in fish biotechnology for the development of many valuable traits to improve the health, well-being, growth rate and other characteristics of fish. Another interesting development in the application of AF(G)Ps is their potential for protection of phospholipid membranes from damage at low, but non-freezing, temperatures. This work is reviewed here as well.

In closing, we express our hope that this monograph will bring a rich new dimension of fish molecular biology to all readers from junior science students to senior researchers. The AF(G)Ps are only one example among the array of adaptations that fish have to their varied and unusual environments. Therefore, in addition to providing essential information on the AF(G)Ps, this work should provide inspiration to researchers who wish to study any new and unusual aspect of fish molecular biology. It is a strange, compelling and fascinating world to discover.

Contents

Chapter 1

Early Research on Proteins from the Antarctic: Antifreeze Glycoproteins

Robert A Feeney and David T Osuga
Food Science and Technology
University of California
Davis, CA 95616, USA

Introduction

Our group's interest in Antarctica began in early 1964, when one of us wrote to an Office of Research, United States Navy, for help in obtaining very fresh penguin eggs.[1] The request was referred to the Office of Polar Programs of the US National Science Foundation. Through the financial and physical help of the Navy and the National Science Foundation, an initial trip was made to Antarctica, on 1 November 1964. This led to seven annual research trips by one or more members of our research group. After the first trip, work was extended to fish proteins, and we were accompanied by students, postdoctoral fellows and colleagues. All told, 20 of us worked there during these seven years. After the first trip, we were accompanied by many others including PhD student Richard G Allison and postdoctoral fellow Herman T Miller in 1965. In 1966, it was student John C Bigler and PhD student Stanley K Komatsu; in 1967, Richard G Allison and Stanley K Komatsu; in 1968, three students, Augusto Trejo Gonzales, James Moore and James Norris, plus two senior co-faculty members, Jerry L Hedrick and Frank E Strong; in 1969, PhD student Jackie R Vandenheede, student Steven Chan, student James Phillips and former PhD student Frank C Greene; in 1970, Jackie Vandenheede, PhD student Ahmed I Ahmed, PhD student CY-K Ho, student Steve Chan and former PhD student Gary E Means. (Then a dozen years later in 1982, two more "went down," including our long-time colleague, Professor Yin Yeh and another student, Michael S Knauf.) Most of the activities during the seven years (1964–1970) have been previously described.[1]

Ross Island

Ross Island, where almost half of our work was done, has been described as one of the most fascinating places in the Antarctica.[2,3] It is located at almost 78° latitude and adjacent to the Ross Ice Shelf in McMurdo Sound. Only 45 miles wide and nearly as long, it has Antarctica's most active volcano, Mt. Erebus, 12,450 feet high and clouded with volcanic vapor. It was once the starting place for famous dashes towards the South Pole.[3] But, the biology at its periphery, including seals, killer whales and penguins, is still attracting scientists as are many different species in the sea. Adelie penguins are at Cape Royds on the western side and at Cape Crozier on the eastern side (nearly 400,000 of them); Emperor Penguins are on the ice shelf also at Crozier, where they are the most southern colony. Anchoring this island of biological splendor is the large base of the US National Science Foundation, McMurdo Station. This is where almost all of the research in Antarctica originates, serving as a "staging area" for other places. Just two or three miles away was the New Zealand's Scott Base, also a research center, and, of course, a jolly place to visit. During late Antarctic spring, airplanes use ice runways, and at least part of each Antarctic summer, McMurdo is served by ships bringing in supplies.

The accessibility of McMurdo Sound by ship made Ross Island a convenient area for early explorers like Robert Scott and Ernest Shakleton to begin their famous dashes to the pole. Their living quarters have been preserved for later visitors to see.[2,3]

Early Studies in Antarctica

Beginning Programs

Our first trip to Antarctica was in November of 1964 to obtain the whites of Adelie penguin (*Pygoscelis adeliae*) eggs.[1] This work required staying in a tent hut for over a week at Cape Crozier, Ross Island. Eggs were opened and the whites were frozen for return to Davis, California within a few hours of their being laid. An additional objective was to determine what laboratory work could be done in the facilities of the US National Science Foundation at the US McMurdo Base on Ross Island, Antarctica.

At McMurdo, we renewed our acquaintance with Professor Donald E Wohlschlag of Stanford University, whom one of us (R.E.F.) had first met a month before during the expedition planning session in Virginia. Professor Wohlschlag was studying the physiological adaptations[4] of fishes living in the icy ocean waters of McMurdo Bay at a temperature of about −1.9°C. Listening to him aroused our interest in the biochemical systems of such creatures living below 0°C. Our previous interests and studies were at very hot temperatures, above 60°C.

Being familiar with the laboratory facilities at McMurdo, we could now plan for research activities the following year in 1965. The laboratory was primarily equipped for biological studies, not chemistry. Therefore, special supplies and equipment had to be brought in by ship during the previous summer. These logistical requirements allowed us to plan only limited laboratory work for 1965. We had to plan for a third year, 1966.

Penguin Egg White and Blood Serum Proteins

In the following Antarctic spring of 1965, our research group was three in number. Our work focused mainly on obtaining a large number of Adelie eggs and freezing the egg whites. We also started work on the penguin blood serum proteins and fish enzymes and fish blood serum proteins.

For the next decade, our laboratories, both in Antarctica and at UC Davis, studied the egg white and blood serum proteins of the Adelie Penguin.[5,6] These studies were developed from long-term research on the properties and evolution of egg white and blood serum proteins.[7] Associated studies were on the egg white proteins of other penguins, particularly those of the Emperor Penguin (*Aptenodytes fosteri*). A

Table 1 Physical and chemical properties of penalbumin, penguin ovalbumin, chicken ovalbumin, penguin serum albumin and bovine serum albumin.*

Property	Penalbumin	Ovalbumin		Serum albumin	
		Penguin	Chicken	Penguin	Bovine
Molecular weight	61,000	48,000	45,000	66,000	66,000
S_{20}, w	3.41	3.25	3.19	4.12	4.40
pI	4.16	3.98	4.48	nd	nd
Carbohydrate (%)	15	7	3	<0.2	0
Phosphorus	0	1	0, 1, 2	0	0
Total (S)	6	6	6	35	35
Sulfhydrls	2	3(4)	4	<1	<1
Disulfide	2	1	1	17	17

*Adapted with permission from Ho *et al.* (1976).[20] nd = not determined.

primary tool in our laboratory had been the use of the extensive differences in the properties of similar proteins in the egg whites of different birds.[7] As compared to chicken egg white, Adelie egg white proteins showed remarkable differences. A major difference was the presence of a large amount of a previously unidentified protein, named *penalbumin* (from the words penguin and albumin), in penguin egg white (Table 1).[6] Biochemical and immunochemical studies of penguin penalbumin have revealed the taxonomic relationship among the penguins and their relationship to other avian species. Upon further investigation, the penalbumin data suggested that their ancestors were capable of flight.[8] Another large difference between the whites was the high amount of sialic acid (~5.0%) in the penguin whites as compared to the chicken's (~0.3%). In contrast, there were less differences between the blood serum proteins of the chicken (*Gallus gallus*) and the penguin.[9]

Fish Enzymes and Blood Serum Proteins

Our interest in the cold-adapted fish increased with a descent into an underwater observation chamber in the ocean. There, 25 feet below the surface of the ice, one could see marine forms everywhere. Our group's interest in the ice-water fish was further generated by encountering Arthur L DeVries, who had wintered over at McMurdo Base. He was working under the auspices of Professor Wohlschlag. DeVries was studying fish caught on hooks from fish houses located over holes cut into the ice above the ocean of McMurdo Sound. With his help, we started on our fish work, collecting samples of blood serum and tissue.

Most of the early studies on enzymes of Antarctic fishes stemmed from the previous studies of Wohlschlag and his students.[4,10,11] Our main studies on Antarctic fish were on the small *Trematomus borchgrevinkii* fish and the large *Dissostichus mawsoni*. Over five years of fishing, approximately 5000 *T. borchgrevinkii* were caught on hooks; about 20 *D. mawsoni* (over 1.3 m long and weighing up to 50 kg) were caught by live capture from seals who brought them into the fish hole to eat (Fig. 1).[5]

Fig. 1 *Dissostichus mawsoni* held by DT Osuga outside fish house on ice of McMurdo Sound, Antarctica, 1966. Photo from DT Osuga.

The activities of six enzymes were studied:[12] muscle fructose-1, 6-diphosphate aldolase, glyceraldehyde-3-phosphate dehydrogenase, glycogen phosphorylase and lactate dehydrogenase, along with the heart mitochondrial cytochrome system and brain tissue acetylcholinesterase.[12] Some important characteristics related to cold adaptation were found, but no generalizations could be made, as cold adaptation was also found in non-Antarctic fish species (e.g. trout).

Many non-scientists, as well as scientists, observed the unusual resistance of polar fish to freezing.[13] The reports of PF Scholander and colleagues, first from the University of Alaska and later from the University of California at San Diego, are seminal.[14–17] They reported that the blood sera of arctic fishes had lower freezing temperatures than those of fishes not adapted to the cold.

Our initial studies on the blood protein of the cold adapted Antarctic fishes concerned a general characterization by limited fractionations, electrophoretic patterns; examination of the effects of temperature on the clotting of whole blood; and on the determination of the level of serum transferrin (an iron-transport protein).[18]

When AL DeVries visited our laboratory in 1967, it was decided that he would join our group sponsored by a National Institutes of Health postdoctoral grant. Supplied with some of the laboratory's Antarctic fish serum, he was helped to complete his study on the substance eventually called the "antifreeze glycoprotein." In early 1968, he joined our laboratory for a period of three years, working with our group on the antifreeze glycoprotein. The main physical and chemical characterization was done in our laboratory and the first paper on the antifreeze glycoproteins was from our laboratory and that of DE Wohlschlag, and co-authored by AL DeVries and DE Wohlschlag.[19]

Early Study of Antifreeze Glycoprotein Structure and Function

Structure

Our laboratory concentrated on the characterization of the antifreeze glycoprotein (AFGP) and its structure, soon resulting in four major publications.[20–23] Several reviews[24–27] summarize much of the early research.

The AFGPs of *T. borchegevinkii* and *D. mawsoni* were separated into eight main components, numbered 1–8, with the larger components represented by the lower numbers, e.g. 8 being the smallest (Table 2).[20] The molecular weights ranged from 2600 Da for AFGP8 to 21,500 Da for AFGP3. A repeating tripeptide of Ala, Ala, Thr, was determined as the stem unit of the longer AFGP 1–5 in which the last residue is glycosidically linked to the disaccharide galactosyl-N-acetylgalactosamine[22,23] (Fig. 2).

The remarkable lowering of the freezing temperature by the Antarctic AFGP[20] is shown in Fig. 3. About 0.6–0.7°C of the freezing point depression in the fish blood is due to the AFGP, with the remainder resulting from colligatively acting substances.[24–39] The effect of the AFGP was considered non-colligative and it generated a thermal hysteresis, which is a difference between the freezing and melting temperatures[24,30] (Table 3).

Functional studies have shown a loss of activity when peptide bonds of the longer AFGP were cleaved, which demonstrated the requirement

Table 2 Chemical composition of freezing point-depressing glycoproteins from *Trematomus borchgrevinki* and *Dissostichus mawsoni*.

	T. borchgrevinki					*D. mawsoni*	
Constituents	3*	4	5	6	8	3, 4, 5	7, 8
	residues/10,000 g						
Threonine	14.6	14.7	14.5	11.4	11.5	14.1	11.0
Alanine	32.5	32.2	32.6	24.9	25.2	31.4	25.0
Proline	0	0	0	3.0	6.5	0	5.0
Galactosamine	14.7	14.4	14.2	14.9	14.0	14.0	13.0
Galactose	17.5	17	17		16.5	17.5	17.0
Acetyl groups		17	16		17	17	16

*Glycoproteins are designated by their electrophoretic band numbers. Reprinted with permission from DeVries *et al.* (1970).[20]

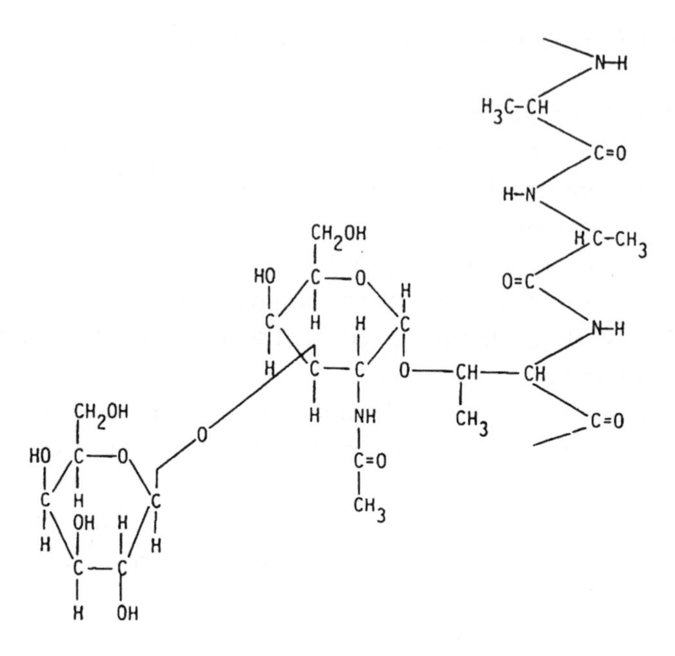

Fig. 2 Polymer unit of AFGP 1–5. Reprinted with permission from Feeney *et al.* (1972).[12]

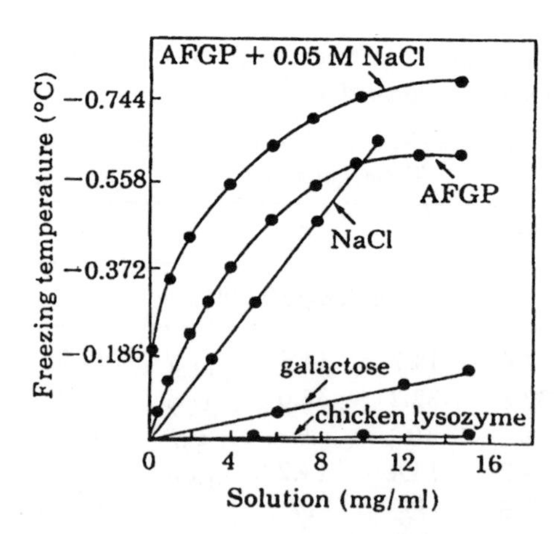

Fig. 3 Freezing points of solutions of sodium chloride, galactose, lysozyme and glycoproteins 3, 4 and 5. A 1 mOsm solution has a freezing point of −0.00186°C. Reprinted with permission from DeVries *et al.* (1970).[20]

Table 3 Freezing and melting of water and solution of antifreeze glycoproteins in water.

Temperature (°C phase)	Observed changes: In water	Observed changes: In water solution of 1% antifreeze
0.0 holding	Melt and freeze	Crystals melt
−0.1 lowering	Frozen	Crystals do not melt, liquid (does not freeze)
−0.7 lowering	Frozen	
−0.8 holding	Frozen	Crystals grow, new crystals form until
−0.7 raising	Frozen	All frozen, no melting
−0.1 raising	Frozen	All frozen, no melting
0.0 holding	Melt and freeze	Melt

Source: Reprinted with permission from Feeney and Roffman (1973).[30]

Table 4 Comparative effects of different treatments* of antifreeze glycoproteins on antifreeze and antilectin activities.

Treatment		Degree of modification (%)	Activity remaining (%)	
No.	Reaction		Antifreeze	Antilectin
1	Oxidation of C-6 alcohols:			
	Galactose	70		
	N-Acetylgalactosamine	15	89	85
2	Aldehyde-bisulfite complex†	(Excess bisulfite)	17	13
3	Oxidation of C-6 to‡:			
	Galactose	70		
	N-Acetylgalactosamine	15	15	12
4	Periodate Oxidation	90	16	11
5	Borate complex	(Excess borate)	10	8
6	Acetylation	32	18	40
7	β-Elimination	80	9	20

*Condition was given in text.
†Bisulfite was added to oxidation product from No. 1.
‡Oxidation of oxidation product of No. 1.
Reprinted with permission from Ahmed *et al.* (1973).[31]

for an intact polypeptide backbone in the AFGP.[20] Acetylation of the carbohydrate side-chains,[20] their β-elimination on alkaline treatment,[18] or formation of their borate complexes[31,32] also caused loss of activity (Table 4).

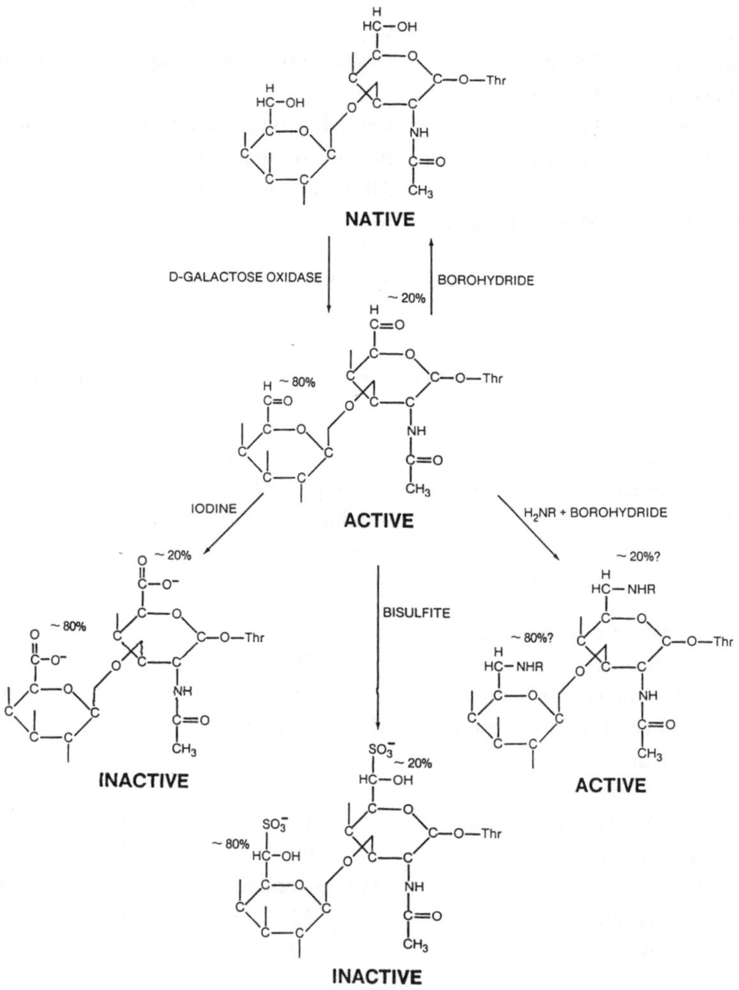

NATIVE

D-GALACTOSE OXIDASE

BOROHYDRIDE

~ 20%

~ 80%

ACTIVE

IODINE

H_2NR + BOROHYDRIDE

~ 20%

~ 80%

INACTIVE

BISULFITE

~ 20%?

~ 80%?

ACTIVE

~ 20%

~ 80%

INACTIVE

Fig. 4 Reaction schemes for oxidations and formation of adduct products of freezing point-depressing glycoproteins. Reprinted with permission from Vandenheede *et al.* (1972).[23]

In contrast, oxidation of the C-terminal hydroxyl groups of AFGP to aldehydes[23] did not inactivate it, but formation of bisulfite adducts or further oxidation of the aldehydes to carboxyls[23] (Fig. 4) did cause losses of activity. Interestingly, the carboxyl derivatives of AFGP showed substantial activity in acidic solutions.[23] Derivatization of the active aldehyde products, by the formation of amide bonds with several different amino acids, also gave products with substantial activity.[33]

Nuclear magnetic resonance (NMR) was a primary tool for characterizing the macromolecular structure of the glycoproteins. The longer structures were suggested to exist in solution as left-handed extended three-fold helices with the disaccharide groups tucked in against the backbone of the polypeptide.[34]

Mechanisms of Function

Many reviews have been published discussing the mechanism of antifreeze function. A few selected ones are by Ananthanarayanan,[35] DeVries,[24,36] Davies and Hew,[37] Davies and Sykes,[38] Duman and DeVries,[28] Feeney,[26] Feeney *et al.*,[39] Feeney and Yeh,[40] Yeh and Feeney,[41] Hew and Yang,[42] and Raymond and DeVries.[29]

A variety of studies then focused on AFGP function.[40] One was the demonstration that mixtures with non-glycosylated AFP from *Pleuronectes americanus* lowered the freezing temperature in an additive fashion, suggesting a similarity in function[43] (Table 5). The early interpretation of function was a direct surface adsorption.[23,28] Different mechanisms have also been suggested, including one affecting the water molecules above the ice surface,[20] and another involving the hydrophobic side-chains, even including clathrate-type inclusion structures.[23] Raymond and DeVries[29] emphasized the model for function in which inhibition occurs by adsorption and the adsorbed molecules raise the curvature of growth steps on the ice surface. This well-depicted "adsorption-inhibition" model still stands, if direct surface adsorption proves to be the function.

Direct evidence for the presence of AFGP near the surface of the ice crystal was obtained by observing the surface second-harmonic

Table 5 Freezing temperatures of *Pleuronectes americanus* antifreeze protein (P.a. AFP), *Boreogadus saida* antifreeze glycoprotein (B.s. AFGP) and their mixture.*

	P.a. AFP		B.s. AFGP		P.a. + B.s. (mg/ml)	
	1	5	1	5	1	5
	Freezing Temperature (°C)					
Experimental	−0.013	−0.158	−0.025	−0.625	−0.045	−0.442
Calculated					−0.038	−0.423
Difference					−0.007	−0.019

*The freezing temperatures of AFGP, AFP and their mixture were determined using a microcuvette (0.2 ml) following the procedure recommended by the instrument instructions; bath temperature of approximately −6°C was used and initiation of crystallization was at −3°C. Reprinted with permission from Osuga *et al.* (1980).[43]

generation intensity of light (532 mμ) at the solution–ice interface.[44] Other evidence included the change in grain boundary curvature in AFGP solutions,[45] the increase in size of gas bubbles extruded in solutions of low concentrations (i.e. 50 μg/ml) of AFGP,[46] and the ice adsorption studies of Knight *et al.*,[47,48] showing the presence of AFGP on specific facets of single ice crystals.

It was observed that the crystal ice structures growing in the presence of AFGP grew needle-like along the c-axis.[29,30] The growth characteristics and kinetics were later described in more detail.[49] When an a-axis oriented crystal was introduced into cold water, c-axis oriented crystals (needle-like) grew at right angles to the original seed crystal.[50]

Early suggestions for the mechanism of lowering the freezing temperature included direct binding at the ice surface, hydrophobic effects and interactions with the water phase.[23–25] This summarized some of our earlier research[1] in Antarctica along with some of the early functional studies on the AFGPs of Antarctic fishes. In particular, our studies of the chemical modifications of AFGPs and their physical interactions with ice, along with the ice crystal growth studies of Knight, DeVries and colleagues initiated a new era in the search to understand the structure and function of these Antarctic glycoproteins.

References

1. Feeney RE (1974). *Professor on the Ice.* Pacific Portals, Davis, CA.
2. Neider C (1974). *Edge of the World, Ross Island, Antarctica.* Doubleday and Company, Inc., Garden City, New York.
3. Feeney RE (1997). *Polar Journeys: The Role of Food and Nutrition in Early Exploration.* University of Alaska Press and American Chemical Society.
4. Wohlschlag DE (1964). Respiratory metabolism and ecological characteristics of some fishes in McMurdo Sound, Antarctica. In: Lee MO (ed.), *Biology of the Antarctic Seas*, Vol. 1. American Geophysics Union, Washington, DC, pp. 33–62.
5. Feeney RE and Osuga DT (1976). Comparative biochemistry of Antarctic proteins. *Comp. Biochem. Physiol.* **54A**: 281–286.
6. Osuga DT, Aminlari M, Ho CY-K, Allison RG and Feeney RE (1983). Sulfhydryl proteins of penguin egg white: ovalbumin and penalbumin. Comparisons with penguin serum albumin, chicken ovalbumin, and bovine serum albumin. *J. Protein Chem.* **2**: 43–62.
7. Feeney RE and Allison RG (1969). *Evolutionary Biochemistry of Proteins Homologous and Analogous Proteins from Avian Egg Whites, Blood Sera, Milk, and Other Substances.* John Wiley and Sons, Inc., New York.
8. Ho CY-K, Prager EM, Osuga DT, Feeney RE and Wilson AC (1976). Penguin evolution: protein comparisons indicate ancestors capable of flight. *J. Mol. Evol.* **8**: 271–282.
9. Allison RG and Feeney RE (1968). Penguin blood serum proteins. *Arch. Biochem. Biophys.* **124**: 548–555.
10. Somero GN and DeVries AL (1967). Temperature tolerance of some Antarctic fishes. *Science* **37**: 72, 257–258.
11. Somero GN, Giese AC and Wohlschlag DI (1968). Cold adaptation of the Antarctic fish *Trematomus bernacchii. Comp. Biochem. Physiol.* **26**: 223–233.
12. Feeney RE, Vandenheede J and Osuga DT (1972). Macromolecules from cold-adapted Antarctic fishes. *Naturwissenschaften* **59**: 22–29.
13. Eliassen E, Leivestad H and Moller D (1960). The effect of low temperatures on the freezing point of plasma and on the potassium/sodium ratio in the muscles of some boreal and subarctic fishes. *Arbok Univ. I Bergen. Mat. Naturv. Serie* No. 14.

14. Scholander PF, Flagg W, Walters V and Irving L (1953). Climatic adaptation in arctic and tropic poikilotherms. *Physiol. Zool.* **26**: 67–92.
15. Scholander PF, van Dam L, Kanwisher JW, Hammel HT and Gordon MS (1957). Supercooling and osmoregulation in arctic fish. *J. Cell. Comp. Physiol.* **49**: 5–24.
16. Gordon MS, Amdur BH and Scholander PF (1962). Freezing resistance in some northern fishes. *Biol. Bull.* **122**: 52–62.
17. Scholander PF and Maggert JE (1971). Supercooling and ice propagation in blood from arctic fishes. *Cryobiology* **8**: 371–374.
18. Komatsu SK, Miller HT, DeVries AL, Osuga DT and Feeney RE (1970). Blood plasma proteins of cold-adapted Antarctic fishes. *Compar. Biochem. Physiol.* **32**: 519–527.
19. DeVries AL and Wohlschlag DE (1969). Freezing resistance in some antarctic fishes. *Science* **163**: 1073–1075.
20. DeVries AL, Komatsu SK and Feeney RE (1970). Chemical and physical properties of freezing point-depressing glycoproteins from Antarctic fishes. *J. Biol. Chem.* **245**: 2901–2908.
21. Komatsu SK, DeVries AL and Feeney RE (1970). Studies of the structure of freezing point-depressing glycoproteins from an Antarctic fish. *J. Biol. Chem.* **245**: 2909–2913.
22. DeVries AL, Vandenheede J and Feeney RE (1971). Primary structure of freezing point-depressing glycoproteins. *J. Biol. Chem.* **246**: 305–308.
23. Vandenheede JR, Ahmed AI and Feeney RE (1972). Structure and role of carbohydrate in freezing point-depressing glycoproteins from an Antarctic fish. *J. Biol. Chem.* **247**: 7885–7889.
24. DeVries AL (1971). Glycoproteins as biological antifreeze agents in Antarctic fishes. *Science* **172**: 1152–1155.
25. Yeh Y and Feeney RE (1996). Antifreeze proteins: structures and mechanisms of function. *Chem. Rev.* **96**: 601–617.
26. Feeney RE (1974). A biological antifreeze. A glycoprotein in the blood of polar fishes lowers the freezing temperature. *American Scientist.* **62**: 712–719.
27. Feeney RE and Brown WD (1974). Plasma proteins in fishes. In: Florkin M and Scheer BT (eds.), *Chemical Zoology*, Vol. 8, *Deuterostomians, Cyclostomes, and Fishes.* Academic Press, Inc., pp. 307–329.

28. Duman JG and DeVries AL (1972). Freezing behaviour of aqueous solutions of glycoproteins from the blood of an Antarctic fish. *Cryobiology* **9**: 469–472.
29. Raymond JA and DeVries AL (1977). Adsorption inhibition as a mechanism of freezing resistance in polar fishes. *Proc. Natl. Acad. Sci. USA* **74**: 2589–2593.
30. Feeney RE and Hofmann R (1973). Depression of freezing point by glycoproteins from an Antarctic fish. *Nature* **243**: 357–359.
31. Ahmed AI, Osuga DT and Feeney RE (1973). Antifreeze glycoprotein from an Antarctic fish: effects of chemical modifications of carbohydrate residues on antifreeze and antilectin activities. *J. Biol. Chem.* **248**: 8524–8527.
32. Ahmed AI, Yeh Y, Osuga DT and Feeney RE (1976). Antifreeze glycoproteins from Antarctic fish. Inactivation by borate. *J. Biol. Chem.* **251**: 3033–3036.
33. Osuga DT, Feather MS, Shah MJ and Feeney RE (1989). Modification of galactose and *N*-acetylgalactosamine residues by oxidation of C-6 hydroxyls to the aldehydes followed by reductive animation: model systems and antifreeze glycoproteins. *J. Protein Chem.* **8**: 519–528.
34. Bush CA and Feeney RE (1986). Conformation of the glycotripeptide repeating unit of antifreeze glycoprotein of polar fish as determined from the fully assigned proton NMR spectrum. *Int. J. Pept. Protein Res.* **28**: 386–397.
35. Ananthanarayanan VA (1989). Antifreeze proteins: structural diversity and mechanism of action. *Life Chem. Reports* 7: 1–32.
36. DeVries AL (1984). Role of glycopeptides and peptides in inhibition of crystallization of water in polar fishes. *Philos. Trans. Roy. Soc. London B* **304**: 575–588.
37. Davies PL and Hew CL (1990). Biochemistry of fish antifreeze proteins. *FASEB J.* **4**: 2460–2468.
38. Davies PL and Sykes BD (1997). Antifreeze proteins. *Curr. Opin. Struct. Biol.* 7: 828–834.
39. Feeney RE, Burcham TS and Yeh Y (1986). Antifreeze glycoproteins from polar fish blood. *Ann. Rev. Biophys. Biophys. Chem.* **15**: 59–78.
40. Feeney RE and Yeh Y (1978). Antifreeze proteins from fish bloods. *Adv. Protein Chem.* **32**: 191–282.

41. Yeh Y and Feeney RE (1978). Anomalous depression of the freezing temperature in a biological system. *Acc. Chem. Res.* **11**: 129–135.
42. Hew CL and Yang DSC (1992). Protein interaction with ice. *Eur. J. Biochem.* **203**: 33–42.
43. Osuga DT, Feeney RE, Yeh Y and Hew CL (1980). Co-functional activities of two different antifreeze proteins: the antifreeze glycoprotein from polar fish and the non-glycoprotein from a Newfoundland fish. *Comp. Biochem. Physiol.* **658**: 403–406.
44. Brown RA, Yeh Y, Burcham TS and Feeney RE (1985). Direct evidence for antifreeze glycoprotein adsorption onto an ice surface. *Biopolymers* **24**: 1265–1270.
45. Kerr WL, Feeney RE, Osuga DT and Reid DS (1985). Interfacial energies between ice and solutions of antifreeze glycoproteins. *Cryo. Lett.* **6**: 371–378.
46. Vesenka JP, Feeney RE and Yeh Y (1993). The microbubble mediated surface probe and the ice antifreeze glycoprotein solution system. *J. Cryst. Growth* **130**(1/2): 67–74.
47. Knight CA, Cheng C-HC, DeVries AL (1991). Adsorption of α-helical antifreeze peptides on specific ice crystal surface plants. *Biophys. J.* **59**: 409–418.
48. Knight CA, DeVries AL and Oolman LD (1984). Fish antifreeze protein and the freezing and recrystallization of ice. *Nature* **308**: 295–296.
49. Yeh Y, Feeney RE, KcKown RL and Warren GJ (1994). Measurement of grain growth in the recrystallization of rapidly frozen solutions of antifreeze glycoproteins. *Biopolymers* **34**(11): 1495–1504.
50. Harrison K, Hallett J, Burcham TS, Feeney RE, Kerr WL and Yeh Y (1987). Ice growth in supercooled solutions of antifreeze glycoprotein. *Nature* **328**: 241–243.

Chapter 2

Physiological Ecology of Antifreeze Proteins — A Northern Perspective

Sally V Goddard and Garth L Fletcher
Ocean Science Centre
Memorial University of Newfoundland
St. John's. Nfld., A1C 5S7, Canada

Introduction

The marine environment off the east coast of Canada is, historically, one of the richest fishing grounds in the world. Upwelling of nutrients on the offshore banks have produced ecosystems teaming with marine life. However, in certain areas, and at certain times of the year, conditions in this region of the world's oceans become extremely inhospitable. During the winter months, icy, subzero conditions prevail. Sea ice can be extensive, and water temperatures regularly approach the freezing point of sea water (about −1.8°C due to the concentration of dissolved salts in the ocean). During the summer, conditions become considerably more clement, and sea water temperatures in general rise well above 0°C. In addition to seasonal variation, environmental conditions vary significantly with latitude and depth.

Animals that are able to colonize areas subject to such environmental extremes are frequently found to have developed survival strategies that match the fluctuating conditions to which they are subjected. We will consider the survival strategies that have developed in a range of teleost fish inhabiting these regions, and will focus on their intriguing ability to produce molecules which act to protect them from freezing in potentially lethal conditions. These are known, collectively, as antifreeze proteins or glycoproteins. In this review, we will discuss the elegant physiological and genetic systems that have developed in northern fish species to produce cold protection matched to environmental demands.

The Problem Faced by Teleost Fish

In the ocean off Canada's east coast, summer water conditions pose no danger of freezing to marine organisms. However, winter conditions should, in theory, be lethal to teleost fish, for the following reason.

The freezing point of any aqueous solution is a colligative property and therefore depends upon the concentration of dissolved solute particles. Distilled water, which contains no solutes, freezes at 0°C. In general, the levels of solutes in teleost body fluids depress the whole fish freezing point to temperatures somewhere between –0.6 and –0.8°C.[1] This means that fish residing in fresh water are in no danger of freezing since the water temperature cannot go below 0°C. Seawater, on the other hand, contains approximately three times the salt concentration of fish blood and freezes at –1.7°C to –2°C, depending on salinity.

Temperatures in the marine environment under consideration regularly fall below –0.8°C, and can reach the freezing point of sea water. In addition, sea ice is a common phenomenon during the winter, and can, depending on latitude, be present well into the spring (Fig. 1). With increasing latitude, there is a general increase in the volume of subzero water present, and because of extended seasonal ice cover, annual insolation is reduced, and waters remain cold for a greater part of the year.[2,3]

Depth also plays a part in the potential lethality of the marine environment during winter. The shallower the water, the greater will be the likelihood of mixing throughout the water column, resulting in very low temperatures from top to bottom, and an increase in the potential for ice crystals to be carried down to the sea bed. Thus, for the winter inhabitants of shallow water, refuges from freezing conditions are few.[4]

Exposure to ice at temperatures below the freezing point of the body fluids has been shown to cause widespread and dramatic

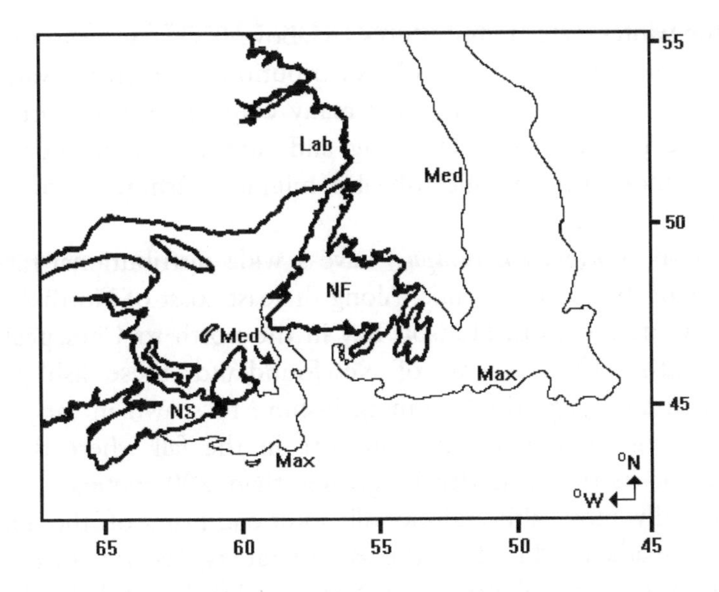

Fig. 1 Median (Med) and Maximum (Max) extent of sea ice off the east coast of Canada in March, for the years 1962–1987. March is generally the month of maximum sea ice cover. Lab = Labrador, NF = the island of Newfoundland and NS = Nova Scotia. Figure modified from Prinkwater *et al.* (1998).[3]

propagation of ice throughout the body fluids of teleost fish, resulting in death.[5–7] Thus, fish inhabiting the northwest Atlantic have, of necessity, evolved a variety of strategies to avoid cold damage and death.

Strategies to Avoid Death by Freezing

Long Distance Migration

Some species avoid the problem of low temperature and ice by carrying out extensive migrations away from areas of danger during the winter. Atlantic salmon (*Salmo salar*), which are found during spring and summer in coastal waters, migrate to warmer waters in winter and may travel great distances offshore, for example to the Labrador sea, the southeastern edge of the Grand Bank, and the west coast of Greenland, where the winter water temperatures are above 0°C.[8,9] Capelin (*Mallotus villosus*), which have a circumpolar distribution in northern waters and provide the major food source for many commercially important fish species, are found inshore in spring and summer where they spawn. However, these little pelagic fish also migrate offshore to overwinter in warmer water.[8]

Lumpfish (*Cyclopterus lumpus*) have a wide distributional range on both sides of the North Atlantic. Along the east coast of North America, they can be found from Hudson Bay in the north to Chesapeake Bay in the south. Off the coast of Newfoundland, these fish perform extensive annual migrations, coming inshore to spawn in early spring, and retreating to deeper water offshore in the fall where they have been caught in trawls at depths greater than 200 meters.

Historically, most, although not all, adult cod living off the northeast coast of Canada would adopt the same strategy, leaving their feeding grounds in the coastal waters of Newfoundland and Labrador, and overwintering considerable distances offshore in deep water. These offshore overwintering concentrations of Atlantic cod formed the basis of a lucrative and ultimately highly destructive, winter fishery on the pre- and post-spawning cod concentrations.[10]

In a reverse strategy, the Arctic char (*Salvelinus alpinus*), a summer resident of marine or brackish water, finds refuge from freezing conditions by swimming upstream and residing in fresh water over winter.[11,12]

Behavioural Adaptations

Some species, such as the longhorn sculpin (*Myoxocephalus octodecemspinosus*), do not perform extensive migrations, but shift location into deeper and generally warmer (0–4°C) nearshore waters for the winter, where they are able to avoid ice crystal contact. The cunner, *Tautogolabrus adspersus,* has developed a very distinctive technique for freeze avoidance in winter. Individuals of this species survive in shallow inshore waters by selecting protected microhabitats such as caves and crevices where they overwinter in a torpid state.[13,14]

The environments these fish occupy are commonly ice covered in winter, with water temperatures well below their plasma freezing point. Thus, the cunners must be surviving in a supercooled state, demonstrating that this is possible, even in shallow, icy waters, provided some effective mechanism exists for keeping ice nucleators out of the body fluids.[15] However, the fact that winter mortalities have been recorded highlights the risk involved in overwintering inshore in Newfoundland.[15] Some flatfishes, such as the winter flounder (*Pleuronectes americanus*), take extra precautions to reduce the risk of ice contact by burrowing into bottom sediments.[16]

While these three species provide examples of behavioural modifications for freeze avoidance, all three have now been shown to have another line of defence against winter freeze damage and death. They are all able to produce antifreeze proteins.[17]

Production of Antifreeze Proteins

The ability of some teleost fish species to produce antifreeze proteins (AFPs) or glycoproteins (AFGPs), is the primary factor that has facilitated their colonisation of very cold waters, and has allowed them to exploit the resources of areas that are periodically ice-laden.[18,19] AF(G)Ps confer freezing protection to fish at low temperatures by binding to nascent ice crystals and .preventing their destructive propagation throughout the body fluids.[20]

The potential presence of AF(G)Ps was first investigated in marine fish collected off the coast of Labrador, by Scholander *et al.* (1957),[7] and subsequently by Gordon *et al.* (1962).[21] They noted that some fish were surviving at depth, supercooled by almost a full degree below the temperature of their surroundings — yet another survival strategy. These fish survived only because of the absence of nucleating ice crystals in their immediate surroundings, so that when brought into contact with ice, they swiftly froze and died. However, other fish species were living in shallow, icy waters and these fish had almost the same plasma freezing points as the sea water they were inhabiting (approximately −1.7°C). These authors termed the solute responsible for lowering the fishes' freezing points, "antifreeze" but did not identify the molecules producing this antifreeze effect. It was not until 1969 that the first AFGPs were identified by DeVries and Wohlschlag (1969).[22]

From that time on, the AF(G)Ps have been the subject of research, and our current knowledge is summarised as follows: (1) in addition to the AFGPs, there are also four different types of AFPs; (2) the five AF(G)P types all have very different molecular structures; (3) evidence suggests that the AF(G)Ps arose by opportunistic amplification of pre-existing gene sequences already coding for proteins with ice-binding potential — a Pandora's box approach to survival which might account for the seemingly random distribution of AF(G)Ps throughout the teleost phylogenetic tree; (4) AF(G)Ps are often produced in the liver and secreted into the blood for systemic freeze protection (liver-type); (5) certain AFPs are also produced in the cells of a variety of tissues throughout the body (primarily peripheral tissues such as the skin or gills) and may serve as a first line of defence against ice damage (skin-type AFPs); (6) in flounder, the genes coding for the liver-type and skin-type are under different control mechanisms, so that the pattern of production of skin- and liver-type AFPs is different within the same fish; and (7) factors (environmental) controlling the presence and levels of AF(G)P vary from species to species, and in some cases, with life history stage.

In the early 1990s, a role for AF(G)Ps in increasing the tolerance of organisms to low, non-freezing conditions (a role not directly related to interaction with ice crystals) was suggested by Rubinsky and colleagues (1991),[23] and Negulescu *et al.* (1992).[24] These authors described a stabilizing effect of AF(G)Ps on mammalian cells when cooled to low temperatures (>0°C), and this has since been attributed to a direct interaction of AF(G)Ps with the cell membrane. Two mechanisms have been suggested to account for this effect. Possibly, by interacting with and blocking ion channels that have become routes for ion leakage at low temperatures, AF(G)Ps help to maintain the integrity of the intracellular environment. This theory is based on the ideas of Hochachka (1986 and 1988)[25–27] that a major cause of cell damage at low temperatures is the decoupling of metabolic activities and membrane function resulting in a passive loss of ion gradients across the cell membrane, a rise in intracellular Ca^{2+} concentration, activation of membrane phospholipid hydrolysis, and cell death. A second mechanism to account for a protective role for AF(G)Ps involves the potential stabilization of cell membranes at very low temperatures by interactions with the phospholipid component to minimize packing irregularities during phase transition. This action has been described in liposomes,[28–30] and in human blood platelets.[31] Evidence that AF(G)Ps might play a role in adaptation and survival of fish in cold, but non-freezing, conditions was provided by Hobbs and Fletcher (1996),[32] and Hobbs (1999)[33] in a study of cold tolerance in goldfish (*Carassius auratus*), in the presence (injected) and absence of AF(G)Ps. However, as yet there is no consensus of opinion about this potentially important role for AF(G)P in adaptation, acclimation and survival of organisms in cold, but non-freezing, conditions.[34]

In spite of the diversity exhibited by the AF(G)Ps, they have one vital thing in common. They have an unusual affinity for ice crystals and are capable of binding to different non-basal surface planes on the crystals. Once bound, they inhibit further ice crystal growth.[17] Antifreeze proteins and glycoproteins are several hundred times more effective at depressing the freezing point than would be predicted based on the number of antifreeze molecules in solution, i.e. their colligative properties.[35,36] However, the unique properties that inhibit ice crystal growth have no implications for melting, and they exert only colligative effects on ice melting points. Thus a diagnostic characteristic of the AF(G)Ps is that their solutions have freezing points lower than their melting points. The gap between these two values is termed a thermal hysteresis and is used as a functional measure of antifreeze activity since it is proportional to the concentration of antifreeze proteins present in solution.[37]

The freeze depressing action of the AF(G)Ps is what enables fish to exploit subzero, icy areas of the ocean for food, reproduction and possibly the avoidance of predators. However, the level of freezing resistance conferred by the antifreeze proteins can differ markedly between species, populations, and life history stages. The hypothesis developed to encompass these facts is that environments characterized by freezing conditions rapidly select for freeze-resistant individuals, and that the pattern of freeze resistance exhibited by a particular population is an indication of the freezing conditions it is likely to encounter during its lifetime under current climatic conditions. It also follows that changes in oceanographic conditions will provoke changes in the antifreeze production profiles of the fish populations affected.

A variety of genetic and physiological mechanisms have been identified, whereby the production of AF(G)Ps can be fine-tuned to match variations in the environment.

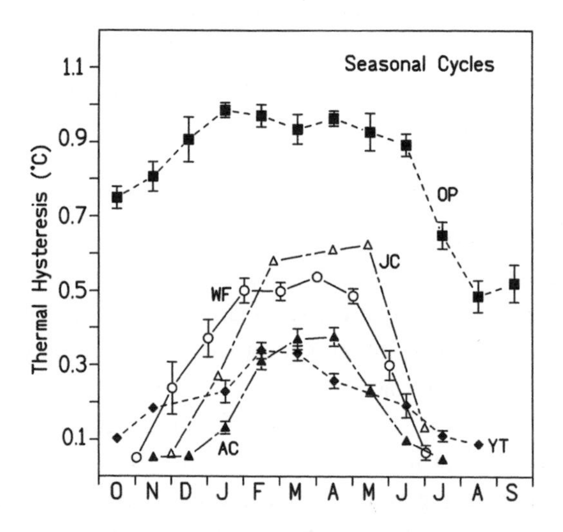

Fig. 2 General representation of seasonal changes in the thermal hysteresis (antifreeze levels) in the blood of several fish species inhabiting Newfoundland waters when exposed to ambient cycles of water temperature and photoperiod. OP = ocean pout (*Macrozoarces americanus*) data re-plotted from Fletcher *et al.* (1984);[66] JC = juvenile Atlantic cod (*Gadus morhua*) data re-plotted from Goddard *et al.* (1999);[69] AC = adult Atlantic cod (*Gadus morhua*) data re-plotted from Rose *et al.* (2000);[50] YT = yellowtail flounder (*Pleuronectes ferrugineus*) data re-plotted from Scott *et al.* (1988);[73] and WF = winter flounder (*Pleuronectes americanus*) thermal hysteresis (TH) in the plasma of fish collected from Chapel's Cove, Newfoundland, held at the Ocean Sciences Centre of Memorial University, and blood sampled monthly throughout the winter (each data point represents the mean ± 1 SE of five to ten fish). All TH values in this figure are means ± 1 SE.

Antifreeze Proteins and Season

Seasonal fluctuations in environmental conditions are frequently mirrored in the physiological responses that they invoke. In sub-polar/boreal/temperate areas of the world's oceans, where seasonal fluctuations can be large, many species of teleost fish have been shown to exhibit seasonal cycles of antifreeze production that mirror the conditions they face in the environment (Fig. 2). In such cases, AF(G)P levels rise to maximum during the winter, usually reaching a peak in February/March. Levels begin to fall during April/May, and reach a baseline during the summer and autumn months. In general, the levels of AF(G)Ps produced and their duration in the plasma, reflect the severity, duration and regularity of freezing conditions that are likely to be encountered by the fish.

Winter Flounder

Winter flounder living in the coastal waters of Newfoundland and Labrador provide a good example of fish that encounter potentially lethal freezing conditions every winter. This species is primarily a shallow water (1–30 meters) resident of the coastal areas of Atlantic Canada and northeastern United States.[12] In Newfoundland, winter flounder are subjected to temperatures that can range from 16°C during late summer to −1.6°C in February and March. During the winter, the inshore areas bear ice of local and Arctic origins. In addition, during storms, ice crystals can be driven deep into the water column.[8] Thus they are obligate producers of antifreeze proteins.

The most obvious environmental cues as to the onset of winter are temperature and photoperiod. Temperatures in the inshore environment are variable from year to year, and can rapidly drop to lethal levels after a comparatively mild December. Thus, it would make sense that winter flounder would use daylength as a signal of the approach of winter because the change in daylength is a more precise indicator of season than water temperature, particularly in shallow areas. This is indeed the case, and the declining daylength experienced during autumn is the major factor controlling the timing of type I AFP appearance in the plasma in November.[38,39] Because of this, all of the fish in a given population are highly synchronous in their annual cycle and there is very little variation from year to year.[40] Production of AFP in the liver of the winter flounder is under the inhibitory control of growth hormone (GH) which indirectly blocks transcription of AFP genes.[41] GH is, in turn, released from the pituitary in a seasonal fashion under the control of the hypothalamic hormones which are regulated by the central nervous system (CNS) in response to photoperiod. Growth hormone releasing factor increases GH secretion and somatostatin decreases it. This complex cascade ensures that AFP mRNA is produced throughout the winter season. However, temperatures must be low enough to ensure stability of the AFP mRNA and allow AFPs to build up in the plasma during winter. Also, clearance of both AFP mRNA and plasma AFP in spring are temperature-dependent. Thus, during a cold spring, antifreeze proteins remain in the blood for longer than usual, while during an unusually warm spring they disappear earlier — a system that provides flexibility in the removal of unnecessary AFPs once the danger of freezing has passed (see review papers Fletcher *et al.* (1985)[17] and Davies *et al.* (1999)[43] and references therein).

The winter flounder is a non-migratory inshore resident that reproduces inshore, and the environment it inhabits is relatively easy to characterize. However, some species have a more complex distribution within the environment, which can vary seasonally, annually, and with life history stage. The Atlantic cod is one such species.

Atlantic Cod

The Atlantic cod (*Gadus morhua*) is one of a number of northern gadoid species known to produce AFGPs in response to low temperatures.[43,44] All the studies on AFGP production described below, have been carried out on adult or juvenile cod caught in waters adjacent to Newfoundland and Labrador.

Experiments with adult Atlantic cod have shown that their seasonal cycles of plasma antifreeze production reflect seasonal temperatures

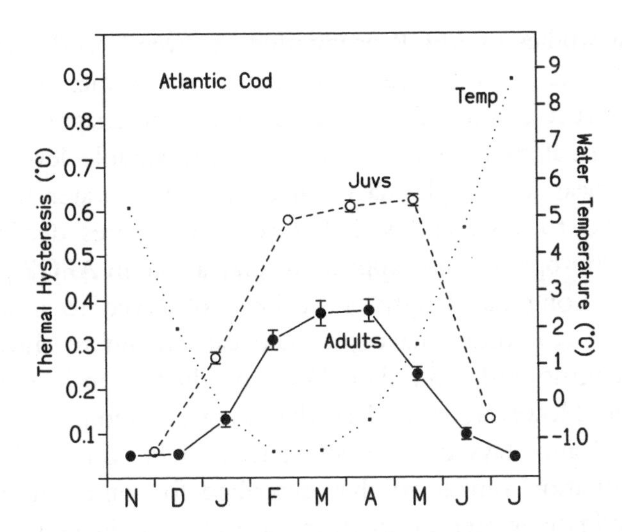

Fig. 3 AFGP production cycles in Atlantic cod (*Gadus morhua*). All adults were collected from coastal waters off the east coast of Newfoundland (principally Conception Bay) between 1981 and 1991 (adults). All juveniles were collected from Conception Bay, Trinity Bay and Notre Dame Bay in the summer of 1992. All animals were held under ambient conditions of water temperature (temp) and photoperiod, and monitored for antifreeze production (plasma thermal hysteresis). Data points are all means ± 1 SE. Figure re-drawn from data contained in Rose *et al.* (2000)[50] (adults), and Goddard *et al.* (1999)[69] (juveniles).

(Fig. 3). AFGPs appear in the plasma in January, a short time (weeks or days) after water temperatures decline below zero (usually during the first week in January).[43,45] Fletcher *et al.* (1980)[43] showed that photoperiod has little effect on AFGP production in adult cod. However, AFGP production can be initiated by exposure to falling temperatures well in advance of winter.[46] Loss of AFGP in the spring is also temperature-dependent, with the biological half-life of AFGP activity being approximately 100 days at 0°C, and 15 days at 5°C.[43] The level of AFGP built up in the plasma during winter is positively correlated with the number of days spent in subzero water temperatures, and AFGP levels reach a maximum after approximately 75 days.[45] Thus, in adult Atlantic cod, initiation of antifreeze production, plasma levels achieved, and clearance times are all controlled primarily by temperature.

The first studies of AFGP production in juvenile cod revealed that, when held over winter under the same environmental conditions, juveniles produce considerably more AFGP and have significantly higher levels of thermal hysteresis than adults.[47] Measurable levels of AFGP begin to appear in the plasma of juvenile cod as water temperatures fall from 2°C to 1°C, and well before temperatures decline to 0°C (Fig. 3).[48] Photoperiod also appears to play a role in AFGP production in juveniles. Long day lengths have been observed to inhibit AFGP production.[49] No studies have yet been carried out to show whether short day lengths could stimulate AFGP production in juveniles during the summer. However, it is clear that as water temperatures decline through 2°C and days become shorter, juvenile cod produce AFGP. This is a situation similar to that described for winter flounder, and suggests obligate winter production of AFGP in juveniles.

These ontogenetic differences in seasonal strategies for AFGP production in adult and juvenile cod from the Newfoundland region reflect their different overwintering environments. Winter distribution of Atlantic cod is influenced by several factors. In addition to their great north-south distribution (see "Antifreeze Proteins and Latitude" below), adult cod are highly mobile, and can undertake extensive seasonal inshore-offshore migrations. Their distribution also depends on their life history stage, and it is stock-specific. It should be noted here that intense fishing pressure on the Northern cod off the coast of Labrador and Northeast Newfoundland (particularly on offshore spawning aggregations), coupled with distributional shifts in the 1980s and 1990s in response to environmental conditions has resulted in a massive decline in numbers; distribution patterns now and in the future are more difficult to predict.[50]

In general, seasonal variation in the distribution of northern cod can be summarized as follows. Adult cod are, by nature, highly migratory. Historically, they have overwintered in great numbers off Labrador and the Northeast Newfoundland Shelf in deep water at temperatures >2°C. These aggregations spawn offshore in spring, and then migrate inshore along highways of warm water to their inshore summer/fall feeding grounds.[51–54] Cod arrive inshore as the water temperatures are warming up. They spend the summer and fall feeding in inshore waters. As water temperatures begin to decline in late fall, they migrate offshore to overwinter once again at their preferred warmer temperatures. Taggart and Frank[55] reported that cod catches in the Northwest Atlantic are generally greatest at a median temperature of 2.9°C. Thus, it would appear that adult cod would generally have little need of antifreeze, since they migrate to avoid freezing conditions.[43] However, this is not always the case. Some adult cod do not migrate offshore, but remain in the inshore regions, and form aggregations that overwinter in deep (>100 meters) subzero water. The cod that remain inshore over winter have a very pressing need to produce antifreeze proteins, and do produce them when exposed to temperature regimes that fall below 0°C.[45] Although these adult cod are found in very cold water, the depths they have been found at are considerably greater than those occupied by the winter flounder, which most likely accounts for their lower thermal hysteresis (mean maximum = 0.46°C) and higher plasma freezing points over winter. At the time of writing, a major concentration of cod off the Northeast coast of Newfoundland is believed to be located inshore, primarily in the Trinity Bay area (see Fig. 7). Water temperatures in this area generally become predominantly subzero between February and March.[4] This highlights the variable nature of adult cod distribution, and the importance of being able to produce AFGP when/if faced with a declining temperature regime.

Juvenile cod are not migratory. Prior to joining the migrating adults (after their third year), it is likely that they exhibit a certain amount of fidelity to their nursery areas.[52,56] The bays of Newfoundland are considered as major nursery grounds for northern cod, and in the first two years of life, catch rates are inversely related to depth.[57] The ability of juvenile cod to overwinter in shallow water inhospitable to many other species may be a survival tactic to avoid being eaten while foraging. Juvenile cod have a good appetite even at subzero temperatures and are capable of eating and growing all winter, if given the opportunity (Brown *et al.* (1989),[58] and Goddard SV — unpublished observations). As cod grow, they tend to move into deeper water, and eventually join the adults at sizes >40 cm, a size that has been reported to be the cutoff point for predation of adult cod on juveniles.[59] Thus, the winter environment inhabited by juvenile cod is shallow, subzero, and ice-covered. Much like the winter flounder, AFGP production and accumulation in the plasma in juvenile cod starts well in advance of the arrival of subzero temperatures, is influenced by photoperiod as well as temperature, and levels of thermal hysteresis achieved are high, and considerably higher than mean maximum levels seen in adult cod.

Thus, while adult cod can be described as opportunistic producers of AFGP with production depending on overwintering strategy, juveniles are obligate producers with a pattern of AFGP production resembling that of the winter flounder. A similar ontogenetic change in antifreeze production has been reported in Atlantic herring (*Clupea harengus harengus*) in the southern Gulf of St. Lawrence. Juvenile herring have high levels of AFP and appear to overwinter in the shallow icy waters of Chaleur Bay whereas adults, with much less AFP, migrate to the deeper ice-free waters of Sydney Bight.[60]

Antifreeze Proteins and Latitude

Many fish species inhabiting the North Atlantic have a large latitudinal distributional range. As a general rule, with increasing latitude, winter conditions in the world's oceans become more severe in terms of low temperature, and area and duration of ice cover (see Fig. 1 for extent of ice cover). Thus, if antifreeze production is matched to the freezing environment encountered, one might expect that in any species with an extensive latitudinal distribution, populations living further/furthest to the north would have developed a greater degree of freeze protection. Studies at both the physiological and genetic levels show that this is, indeed, the case for several species.

As previously discussed, environmental factors (photoperiod and temperature) influence the timing and level of antifreeze produced in the plasma. Thus, in order to identify real population differences in antifreeze production capacity within a species, it is important to subject all groups to the same environmental stimuli. In practical terms, this entails moving groups of fish to a common location and following their antifreeze production over a winter cycle. Winter flounder, ocean pout and Atlantic cod have been studied in this manner. In each case, there were genetic differences in AF(G)P production cycles that could be matched to environmental demands.

Winter Flounder

Comparisons of antifreeze production have been made between winter flounder from (north to south) Conception Bay, Newfoundland, St. Margaret's Bay, Nova Scotia, and Passamaquoddy Bay, New Brunswick, Canada [Fletcher *et al.* (1998),[61] and Fletcher GL, unpublished data]. These studies have shown that while Newfoundland flounder produce the highest levels, production appears similar in flounder from Nova Scotia and New Brunswick, most likely reflecting the severity of the environment. Similarly, timing of production reflects the onset and duration of severe winter conditions. Flounder from New Brunswick develop measurable levels of plasma approximately two months later than flounder in Newfoundland and start to lose their antifreeze about one month earlier. These observations correlate with the fact that water temperatures are generally warmer throughout the winter in Passamaquoddy Bay than they are in Newfoundland, and that water cools to subzero temperatures considerably later, and warms up earlier in spring. This difference in timing seen in field studies simply cannot be attributed to differences in water temperature and photoperiod, because when the New Brunswick flounder were transported to Newfoundland and maintained for up to two years in the same aquaria as Newfoundland flounder, they retained the timing of their New Brunswick antifreeze cycle (Fig. 4), (Fletcher GL, unpublished data). Similarly, flounder from Nova Scotian waters held in Newfoundland begin to accumulate plasma antifreeze approximately one month later, and lose it two months earlier than Newfoundland flounder.[61] We suggest that the timing of the annual antifreeze cycle is a genetically determined

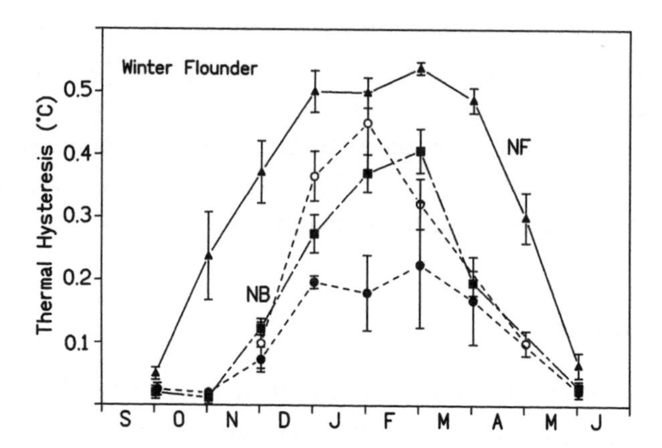

Fig. 4 Annual cycle of blood plasma thermal hysteresis in Newfoundland (NF) and New Brunswick (NB) winter flounder, showing population differences in onset of production (retained even after two years in NF), and antifreeze production capacity. NB winter flounder were collected from Passamaquoddy Bay, NB, and transported to NF, where they were maintained in aquaria along with the NF flounder. Two groups of NB winter flounder were studied in this experiment. The first group was transported in September and held under NF conditions for two winter cycles (open circles = first winter, and closed circles = second winter). The second group was transported to NF one year after the first group, again in September. Timing of the annual antifreeze cycle in the NB flounder is very similar to that observed in NB flounder sampled in the field at monthly intervals from Passamaquoddy Bay NB.[62] All TH values are means ± 1 SE of five to ten fish.

population characteristic which is the result of environmental selection.[61,62]

The exception to the rule that reminds us to be cautious when we use latitude as a proxy for environmental severity, is a field study of antifreeze production in winter flounder from Newfoundland, New Brunswick, and Shinnecock Bay, Long Island, New York.[62] In this study, the Newfoundland and New Brunswick flounder field antifreeze cycles met expectations based on previous investigations. However, the Long Island flounder (the most southerly population examined) started to produce antifreeze at the same time as the most northerly (Newfoundland) population. An investigation of the environment inhabited revealed that Shinnecock Bay is a very shallow, enclosed bay, being kept open to the ocean only by dredging procedures. Water temperatures in the bay are strongly affected by air temperatures. Temperatures can fall rapidly with lowest water temperatures occurring in January/Febuary. Input of fresh water (river runoff and seepage of groundwater) results in lower salinities, higher seawater freezing points, and more ice than might be expected. Considerable amounts of ice can be present in January.[62] Therefore, the Shinnecock Bay flounder need to be prepared for freezing conditions at the same time as the Newfoundland flounder in order to avoid freezing in their icy shallow water location. However, air temperatures warm up well in advance of conditions in Newfoundland, and this is reflected in the timing of loss of antifreeze, which more closely resembles that for New Brunswick and Nova Scotia flounder. Thus, a careful study of the environmental conditions makes sense of the antifreeze production profiles.

This caveat must also be remembered when considering the antifreeze gene complement of populations from different areas. The presence of ice and severity of overwintering conditions are generally, but not invariably, correlated with latitude. An understanding of the severity of local winter conditions is important in understanding both the pattern of antifreeze production, and also the antifreeze gene copy number and arrangement within the genome. Probes are available for the detection of AFP genes in the winter flounder genome. In populations subjected to ongoing selection pressure one might expect to see a uniform pattern at the genetic level showing high copy number and an amplified ability to produce antifreeze proteins. However, in populations where selection pressure has been relaxed, it might be prudent to have no fixed ideas about what to expect at the genetic level, since in the absence of selection there might be more variation within the population. A study of population differences in AFP gene copy number and arrangement in winter flounder from nine widely distributed locations along the east coast of Canada, the Gulf of Maine and Long Island showed that this is the case.[63] AFP genes in the winter flounder (and there are a lot of them) exist as linked but irregularly spaced genes, and also as a series of direct tandem repeats. The tandemly repeated genes are thought to be the result of gene amplification events in response to intense selective pressure, and provide a method for the rapid stepping up of AFP levels.[42,64] Hayes and colleagues found that while flounder from the most environmentally challenging areas did have high AFP gene copy numbers and large tandem components, it was difficult to draw any conclusions about fish from the other sites. The authors concluded that gene amplification is the result of selection pressure, resulting in higher plasma levels of AFP and the ability to withstand extreme conditions of low temperature and ice during winter. However, if climatic changes result in a relaxation of that selection pressure, then it might be reasonable to expect considerable variation within a population in terms of AFP gene dosage and levels of expression. This variation in gene family size and organization was seen in the winter flounder populations studied by Hayes *et al.* (1991).[63] It is also evident in the phenotypic expression of AFP in a population of sea raven from Passamaquoddy Bay (New Brunswick), the location of one of the winter flounder populations studied by Hayes *et al.*[63]

Sea Raven

The sea raven is an exception among the species studied to date in that it exhibits large individual variation in AFP levels within a population. Field studies of a population in Passamaquoddy Bay, New Brunswick, revealed that approximately half of the fish had low blood antifreeze levels, and of the remaining 50%, a few individuals possessed sufficiently high antifreeze concentrations to protect them down to the freezing temperature of seawater.[65] An illustration of the range of sea raven blood plasma freezing temperatures is presented in Fig. 5. These observations suggest that high levels of antifreeze proteins in the blood may have been essential at one time, but are not necessary for survival of sea raven in Passamaquoddy Bay at the present time. However if winter conditions should become more severe, there are individuals within the population that could survive in the presence of icy seawater.

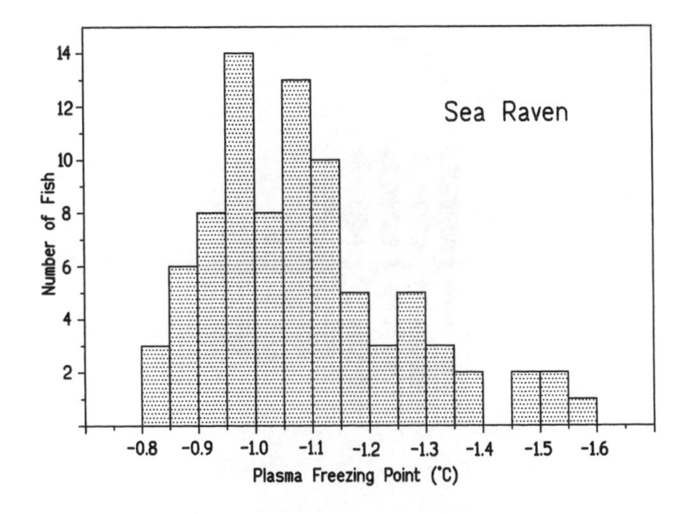

Fig. 5 Frequency distribution of blood plasma freezing points in sea raven sampled from Passamaquoddy Bay, New Brunswick, during the winter months (Febuary to May); N = 91. Data re-plotted from Fletcher *et al.* (1986).[65]

Ocean Pout

The most straightforward relationship between environment, AFP levels and gene copy number can be seen in studies comparing ocean pout from Newfoundland and those from Passamaquoddy Bay, New Brunswick (Fig. 6).[66] A dramatic difference is evident in antifreeze production capacity between the two groups, so that during the winter months, the Newfoundland ocean pout would be considerably more freeze-resistant (−1.7°C) than their New Brunswick counterparts (−1.0°C). Subzero water temperatures occur considerably less frequently in Passamaquoddy Bay than they do in Newfoundland. Therefore, New Brunswick ocean pout may have little or no need for AFPs.

When the AFP gene families of these two groups of fish were compared by genomic Southern blotting, it was estimated that Newfoundland ocean pout contained approximately 150 copies of the AFP gene, whereas New Brunswick ocean pout possessed less than one quarter of that number (Fig. 6).[67] They also noted that the ocean

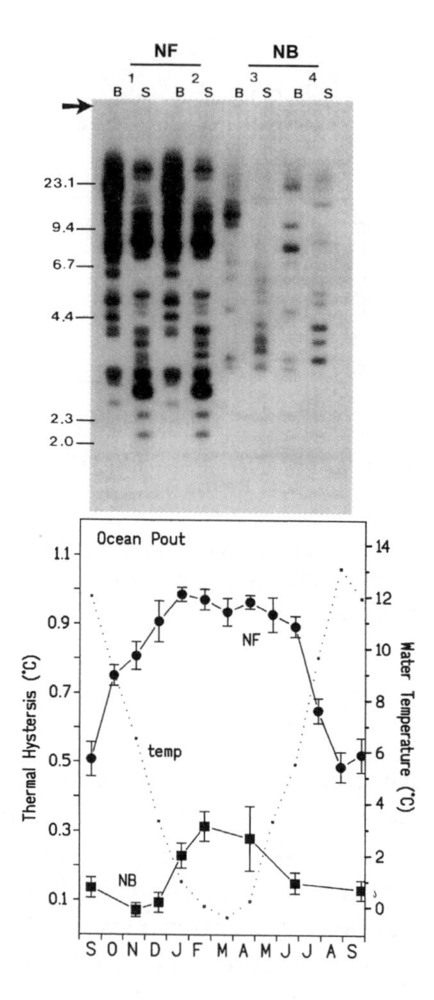

Fig. 6 Ocean pout AFP genes: population differences in the multigene family and AFP production. Upper panel: Southern blot of testes DNA from two Newfoundland (NF) and two Passamaquoddy Bay, New Brunswick (NB) ocean pout, digested with Bam H1 (B) or Sst (S), and probed with a cloned ocean pout AFP cDNA.[67] Arrow indicates the origin. Numerals on the left indicate DNA size fragments (kilobases). Lower panel: Seasonal cycles of blood plasma freezing points in Newfoundland (NF) and New Brunswick (NB) ocean pout held in aquaria in Newfoundland under Newfoundland conditions of temperature and photoperiod. Values are expressed as means ± 1 SE. Data re-plotted from Fletcher *et al.* (1984).[66]

pout AFP genes resembled the irregularly spaced components of the winter flounder gene complement, rather than the tandem repeats which are the sites of choice for AFP gene amplification in the winter flounder. Southern blotting indicated that one population profile could not have been generated from the other by either straightforward deletion or duplication processes. The authors concluded that gene amplification events had occurred in the more northerly population.[67] Thus, there appears to be a strong positive correlation between AFP gene dosage, protein levels and severity of the overwintering environment.

Atlantic Cod

Atlantic cod are widely distributed on both sides of the North Atlantic. To the east, they are found from the Barents Sea and around Iceland, in the Baltic Sea and as far south as the Bay of Biscay. Cod in the northwest Atlantic historically cover a wide latitudinal range, from Ungava Bay, Baffin Island and Greenland in the north, as far south as Cape Hatteras N.C. lat 35°10′ N.[68] Thus, it seems likely that there will be differences in the AFGP production capacity between populations, and that these will be related to the winter environment inhabited.

Studies on AFGP production have been carried out on Atlantic cod collected from a variety of sites around Newfoundland, and from Nova Scotia and New Brunswick, and held in common environments at the Ocean Sciences Centre of Memorial University in Newfoundland (Fig. 7). Cod collected from Newfoundland sites tend to outperform those from further south (Nova Scotia and New Brunswick) (Goddard SV — unpublished data). However, the most conclusive evidence that cod from more extreme environments have greater antifreeze production capacity comes from two studies with northern cod, one on juveniles and the other on adults.[50,69]

AFGP production capacity was investigated in juvenile cod collected from four locations along the northeast coast of Newfoundland and held together in a large tank over the winter (Fig. 7).[69] The population from the most northerly site (St. Lunaire at the tip of Newfoundland's

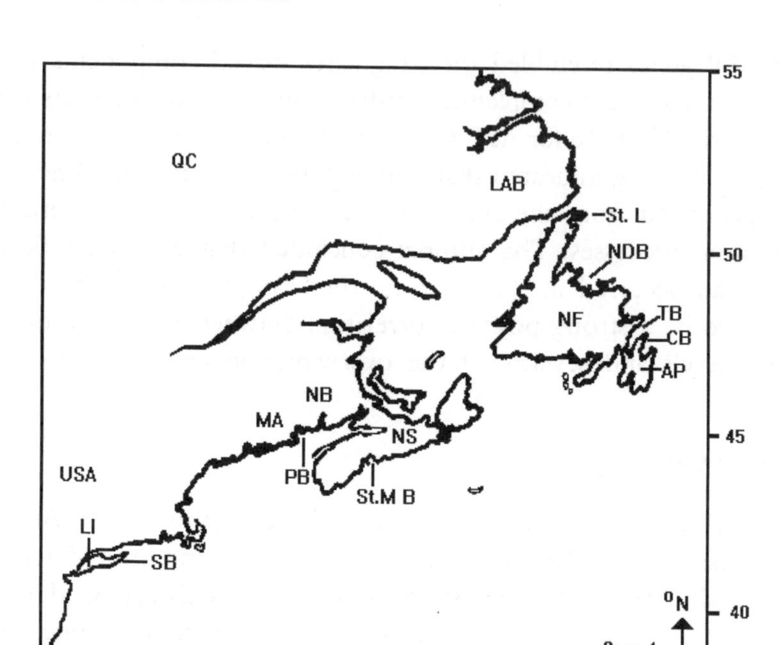

Fig. 7 Locations of sites mentioned in the text, where various fish species have been collected for further laboratory studies, or have had blood samples collected from them in the field. LAB = Labrador, St.L = St. Lunaire, NDB = Notre Dame Bay, TB = Trinity Bay, CB = Conception Bay, AP = Avalon Peninsula, NF = Newfoundland, NS = Nova Scotia, St. MB = St. Margaret's Bay, PB = Passamaquoddy Bay, NB = New Brunswick, QC = Quebec, LI = Long Island, and SB = Shinnecock Bay.

Great Northern Peninsula), showed augmentation of the two features characteristic of obligate antifreeze production — early onset of production and greater levels of thermal hysteresis throughout the winter. Cod from the three other locations all produced high levels of AFGP, and commenced production early in the winter, as expected. However, the St. Lunaire juveniles had very high AFGP levels by early January, and out did even the winter flounder in the degree of freeze protection developed by mid-winter (Fig. 8). Evidently, all the populations inhabiting the northeast coastal regions of Newfoundland

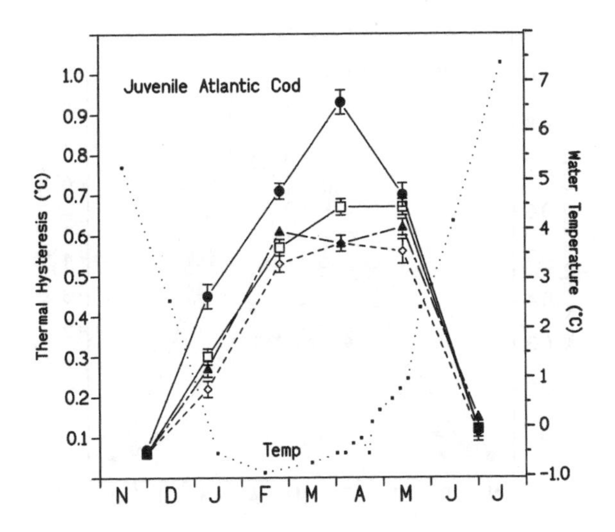

Fig. 8 Cycles of AFGP production (TH = thermal hysteresis) in four groups of juvenile Atlantic cod (*Gadus morhua*) collected from St. Lunaire, at the tip of Newfoundland's Great Northern Peninsula (filled circles, solid lines), Notre Dame Bay (filed triangles, mixed dashed lines), Trinity Bay (open diamonds, dashed lines), and Conception Bay (open squares, solid lines). Cod were all held together in one tank at the OSC, Logy Bay, and monitored for plasma TH throughout the winter and spring months. All values of TH are means ± 1 SE. Data re-plotted from Goddard *et al.* (1999).[69]

have been selected to be well equipped to deal with extreme winter environments characterized by subzero temperatures and ice. Those from the highest latitudes have experienced even greater selective pressures due to environmental factors such as earlier onset and later retreat of ice cover, reduced annual insolation, extended period of exposure to subzero temperatures, and increased chances of vertical mixing of ice crystals down through the water column.[69] This is reflected in their AFGP production.

AFGP gene dosage and gene organization have not been investigated in the genome of these different cod populations. Thus, it is not possible to ascertain whether population differences are due to differential survival of the most cold-resistant individuals from a whole suite of cold tolerances within a panmictic population, or whether, as

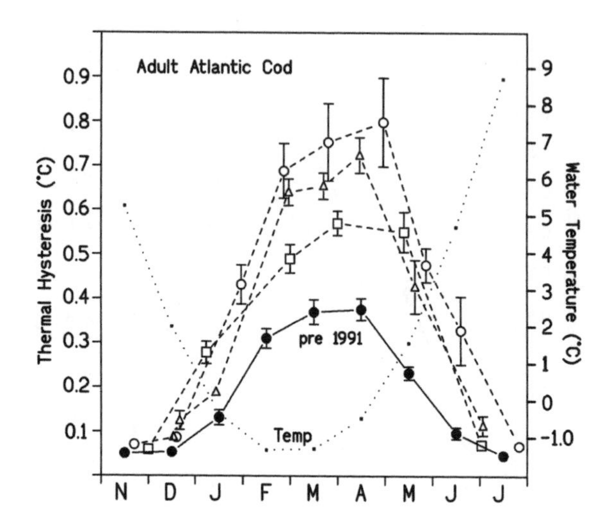

Fig. 9 Abrupt shift in winter AFGP cycles between adult cod caught off the east coast of Newfoundland prior to the winter of 1990–1991, and cod caught in subsequent years. Pre-1991 graph (closed circles, solid lines) = the mean of thermal hysteresis (TH) cycles of adult Atlantic cod caught off the east coast of Newfoundland in the summers of 1986, 1987, 1988, 1989 and 1990, and held and monitored for plasma TH during each subsequent winter. Graphs with open symbols and dashed lines represent plasma TH levels of adult Atlantic cod caught off the east coast of Newfoundland in summer 1991 (circles), 1992 (squares) and 1994 (triangles), held in aquaria, and monitored during the subsequent winter [data modified from Rose *et al.* (2000).[50]] Water temperature graph = means of holding tank temperatures taken a minimum of twice per week during the winters of 1987–88, 1988–89, 1989–90, 1990–91, 1991–92 and 1992–92.

seen in the ocean pout, AFGP gene amplification as a result of environmental pressure has taken place in the Atlantic cod. Based on the very high levels of thermal hysteresis seen in the St. Lunaire juveniles, levels that were not even represented in the other three populations (comprised of over 100 animals), we believe that the latter explanation is the correct one. A further study on adult cod highlights the importance of research into this question.

A long-term (>ten years) database of AFGP production patterns in cod caught off the Avalon Peninsula of Newfoundland revealed that in 1991, a group of cod with much greater AFGP production capacity

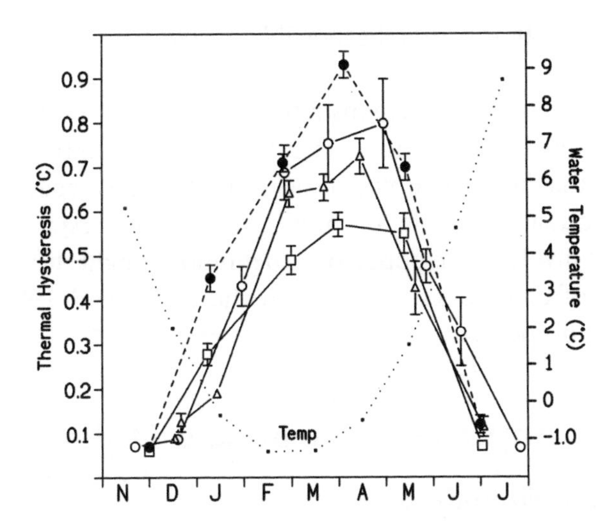

Fig. 10 Comparison of post-shift (northerly) adult Atlantic cod [open symbols and solid lines; circles = 1991–92, squares = 1992–93, triangles = 1994–95 (see Fig. 9)] with northerly juveniles collected from St. Lunaire (Great Northern Peninsula). Data re-plotted from Rose *et al.* (2000)[50] (adults) and Goddard *et al.* (1999)[69] (juveniles).

were present in the area (Fig. 9). Representatives of this group were still in evidence in 1994.[50] These cod are believed to be southerly migrants from more northerly areas and, possibly, the remnants of a northerly population physiologically specialized to live at higher latitudes. Comparison of AFGP production patterns seen in the northerly adults and the St. Lunaire juveniles lend weight to this conclusion (Fig. 10).

Evidence of physiological diversity in a fish species that is subject to heavy fishing pressure is of great importance. In the past, cod off the coast of Newfoundland and Labrador were managed as a single stock complex. Little consideration was given to the possibility that fragile populations especially well adapted to survive in certain areas might need more specialized management plans to ensure their sustainability and even their survival. As fishing pressure increased on cod off the east coast of Canada, the cod fisheries off the coast of North Labrador were the first to collapse. They still show no signs of recovery.[50,70] It may not be possible for cod from more southerly populations to re-seed these areas. These studies of AFGP production capacity in Atlantic cod demonstrate the need for understanding of the complexities of population structure before exploiting any natural resource to the limit. The cod fishery off the East coast of Canada provides a classic example of how a better understanding of the complexities involved in population definition might have prevented the collapse of cod stocks witnessed in the 1980s and 1990s and still being endured today.

Antifreeze Proteins and Depth

The Righteye Flounders

As noted earlier, it is the combination of supercooling and ice contact that is lethal to fish. Therefore, even if fish do reside at temperatures below their freezing points they are in no danger of freezing as long as they avoid contact with ice. Since ice can only be propagated into the water column from the surface, fish can avoid freezing by residing in waters at sufficient depth to avoid ice contact.

There is little doubt that environments characterized by freezing conditions select for freeze-resistant species. Therefore, it is reasonable to hypothesize that the level of freezing resistance exhibited by a species or a particular population is a reflection of the freezing conditions they are likely to encounter. This hypothesis was examined with respect to seven species of righteye flounders inhabiting Newfoundland waters, namely the Greenland halibut or turbot (*Reinhardtius hippoglossoides*), the halibut (*Hippoglossus hippoglossus*), the witch flounder, or greysole (*Glyptocephalus cynoglossus*), the American plaice (*Hippoglossoides platessoides*), the yellowtail flounder *(Pleuronectes ferrugineus)*, the winter flounder, and the smooth flounder *(Pleuronectes putnami)*. The environments occupied by these seven species off the coast of Newfoundland and Labrador fall into three approximate groupings (Fig. 11).

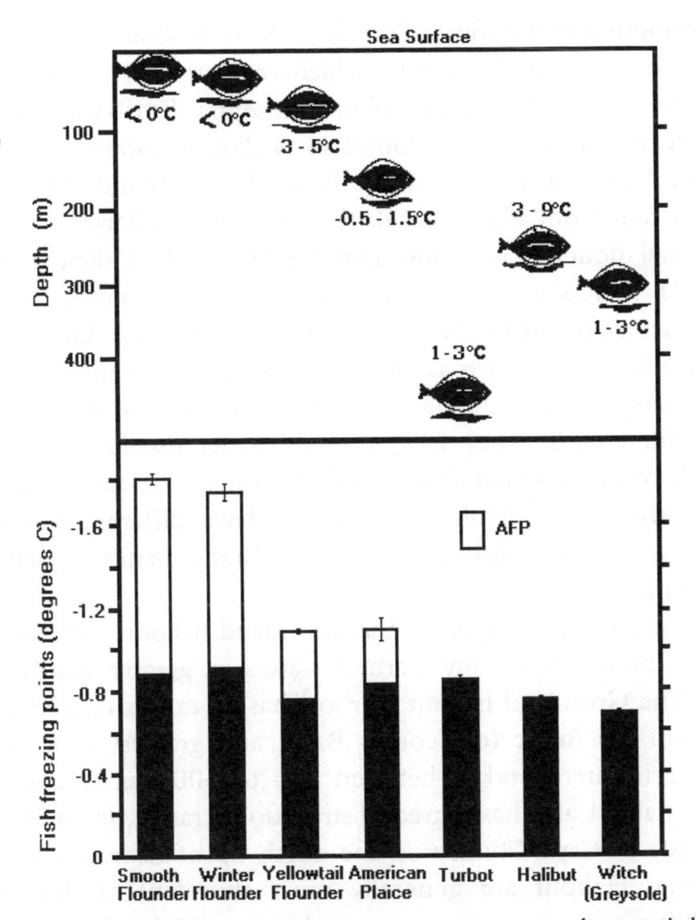

Fig. 11 Depth distribution, water temperatures most commonly occupied in winter (upper panel), and plasma freezing points (lower panel) of the seven species Righteye flounders inhabiting Newfoundland waters. Total plasma freezing points are plotted as means ± 1 SE. The clear upper areas of the smooth flounder, winter flounder, yellowtail flounder and American plaice histograms represent the AFP contribution to the plasma freezing points. Numbers of blood samples analyzed for each species were as follows: smooth flounder, n = 13; winter flounder, n = 20; yellowtail flounder, n = 10; American plaice, n = 12; turbot, n = 20; halibut, n = 10; and witch (greysole), n = 19. Data for American plaice, turbot, halibut and witch were obtained from blood samples collected during a March groundfish survey by the Department of Fisheries and Oceans, Canada, on the Grand Banks. Smooth flounder were collected from Passamaquoddy Bay, New Brunswick. Yellowtail and winter flounder were caught in Conception Bay, Newfoundland. All samples were collected in March.

The smooth and the winter flounder inhabit shallow waters of depths considerably less than 40 meters, which regularly fall below zero and become ice covered in winter. Their latitudinal distribution is similar, with smooth flounder being found in shallow coastal/estuarine areas along the Labrador coast and south to Rhode Island, while winter flounder occur from Hamilton Bank (off Newfoundland) to Georgia.

Yellowtail flounder are most commonly found at depths between 37 and 91 meters and in a preferred temperature range of 3°C to 5°C. Their most northerly distribution is over the Grand Banks. American plaice are known as cold water fish, and are most commonly caught at temperatures ranging from –0.5°C to 1.5°C at depths between 125 to 200 meters. They have a much wider distribution than the yellowtail, being reported from Baffin Island in the north, to Rhode Island in the south. While these species have different temperature preferences, they are both found on the Grand Banks at depths less than 200 meters.

The remaining three species, the Greenland halibut, the halibut and the witch, prefer to occupy warmer waters at greater depths (> 200 meters). The Greenland halibut (turbot) has an extensive distributional range, from the Arctic to Georges Bank, and greatest concentrations of this species are found at between 275 to 600 meters and 1°C to 3°C. The halibut also has a great distributional range, from Disko Bay (Greenland) to Cape Chidley on the north tip of Labrador and south to Virginia. Halibut are generally most abundant at depths from 200 to 300 meters and at temperatures between 3°C and 9°C, although in winter, their preferred depth range is between 200 to 500 meters in deep water channels off the edge of the continental shelf. Witch flounder can be found from Hamilton Bank to Cheaspeake Bay and occupy a similar niche to the Greenland halibut. They are most abundant at depths between 200 to 400 meters and temperatures between 2°C to 6°C. All data on depth and temperature preferences and distributional ranges of righteye flounders, both in the text and in Fig. 11, have been abstracted from references Pinhorn (1976)[10] and Scott (1988)[12] Canadian Department of Fisheries and Oceans Science Stock Status Reports for the species discussed, and Canadian

Department of Fisheries and Oceans Underwater World Information Series for the species discussed.

A mid-winter comparison of AFP levels in the plasma of these seven species of righteye flounder reveals a correlation between depth distribution and freeze protection. The three species of righteye flounders inhabiting the deeper (>200 meters) waters off the edges of the continental shelf (Greenland halibut, the halibut and the witch flounder) were caught and sampled directly from the wild during a March groundfish survey being carried out by the Canadian Department of Fisheries and Oceans (Fletcher GL, unpublished data). No AFPs were detected in the plasma of these fish, and their plasma freezing points ranged from –0.7°C to –0.85°C (Fig. 11). The American plaice and yellowtail flounder inhabit somewhat shallower waters of the Grand Banks (<200 meters). When these species were sampled in the wild, they were found to possess moderate AFP levels that would protect them from freezing down to temperatures of approximately –1.1°C. The species occupying shallow inshore waters (winter flounder and smooth flounder) showed the highest concentrations of AFPs (Fig. 11). These species are capable of surviving in icy waters at temperatures approaching the freezing point of seawater (–1.7°C to –1.8°C). It is reasonable to assume that by mid-winter (March), fish in the wild will have developed a level of thermal hysteresis sufficient for winter survival, and that this is representative of their AFP-producing capacity. How is this phenotypic expression of AFP mirrored in AFP gene complement?

AFP genes responsible for production of AFP in the plasma of the winter flounder have received most attention and can be summarized as consisting of 30 to 40 members, some of which are arrayed in direct tandem repeats, while the remainder are linked, but irregularly spaced.[64,71] Scott *et al.* (1986)[72] speculate that patterns of gene amplification may be the signposts to understanding antifreeze evolution; tandem amplification in the winter flounder has been put forward as the method by which a rapid increase in AFP output was achieved, allowing adaptation to cooling environments and exploitation of shallow water resources.[64,72] Using the winter flounder as a benchmark, the AFP gene complement of several other righteye flounders has been examined.

AFP gene patterns in flounders representative of the three depth groupings (shallow, mid-water and deep) correlate well with expectations based on AFP levels and habitat. Southern blots of genomic DNA from the two shallow water species (winter flounder and smooth flounder) show strong similarities when probed with a winter flounder AFP cDNA clone. Tandemly iterated components are present in both genomes suggesting rapid adaptation to extreme freezing environments. Whether tandem amplification took place before or after speciation is a matter for speculation. However, Scott *et al.* (1988)[73] note that the lengths of restriction fragments from the tandem repeats are quite different for the two species, suggesting amplification after speciation.

Yellowtail flounder and American plaice both have depth preferences <200 meters, and their distributions overlap on the Grand Banks. Their plasma antifreeze levels measured in the field in mid-winter were very similar: and at the genetic level, the American plaice displays an AFP gene hybridization pattern qualitatively similar to the yellowtail flounder.[73] Neither of these species possess the tandem repeat component found in the species occupying shallow water, although they possess AFP genes homologous to the irregularly spaced winter flounder AFP genes.[72,73] An analysis of AFP gene copy number for winter flounder and yellowtail flounder reveals a ratio of 3:1, and this is consistent with protein levels in the plasma when considered on a molar basis.[73] So, while these species develop some additional freeze protection in winter, it is significantly less than that required by the shallow water species.

None of the representatives of the three species inhabiting deep water had any AFPs in their plasma when blood sampled in the field in March, consistent with the fact that the environments they occupy pose no danger of freezing. A search for AFP genes in witch flounder genomic DNA revealed a weak signal when probed using a winter flounder AFP genomic clone, possibly indicative of the presence of AFP genes homologous to the winter flounder irregularly spaced AFP component.[72] Might this signal represent a gene sequence coding for a protein with an unexploited potential for ice binding — an AFP progenitor? Might it represent remnants of a higher AFP gene complement, lost over time? Or might the witch flounder possess AFP genes that have diverged in sequence over time to the point where only weak hybridization is to be expected? Intuitively, the first two options seem most likely. However, the appropriateness of the probe being used must be considered before drawing too many conclusions. This may be a factor in the lack of signal strength seen in the American plaice blot when probed with a winter flounder AFP cDNA clone resulting in qualitative, but not quantitative similarity with the yellowtail flounder blot (Scott *et al.* (1988),[73] Davies PL and Gauthier S, personal communication). The development and use of different probes, and a search for AFP genes in other representatives of this deep water group might help to answer this question.

The correlation between gene amplification events, AFP levels and depth distribution in the righteye flounders is striking. The more susceptible a species is to freezing, the deeper its habitat tends to be. Conversely, the deeper the habitat, the less the requirement for freeze protection. The obligate AFP producers living in shallow water have both the irregularly spaced and tandemly iterated AFP gene components. Those living in the mid-depth range show only the irregularly spaced component, while the DNA of the one representative of the deep water flounders examined showed only a faint AFP gene signal. From these data, it is apparent that the deep water flounders are in no danger of ice crystal contact, while the shallow water flounders face this danger every winter: but what is the risk to the group inhabiting the mid-depth range?

If the presence of antifreeze peptides in American plaice and yellowtail flounder is the result of environmental selection, these fish populations must, on occasion, be exposed to freezing conditions. However, in order for benthic fish to freeze, ice crystals would have to be transported from the surface to the bottom. Can this occur in 60 to 200 meters of water?

Mechanisms can be suggested by which ice crystals might be carried down to the sea bed over the Grand Banks. Wind driven turbulence mixes the upper water column during periods of intense winds. If strong winds were to occur together with surface ice formation, and minimum density stratification of the water column, the resulting turbulent mixing might take ice crystals to the bottom. Convective overturning as a result of a density inversion could also be postulated as a method by which ice crystals might be carried to the bottom. This case scenario requires cooling and evaporation at the surface in the presence of ice, with resulting increase in surface density at a time when the entire lower water column is approximately uniform in temperature and salinity (density). This would result in instability in the water column due to a density inversion — more dense water sitting on top of less dense water — resulting in a rapid overturning of the surface and deep layers, transporting the more dense, ice-laden surface water to the bottom. Conditions favouring these two methods (strong winds, low temperature, low density stratification of the water column, ice and very low temperatures at the surface) are most likely to occur between February and April off the east coast of Newfoundland and Labrador.[8,74–76] Buoyancy and melting rate of ice crystals would also have to be considered. As far as we can determine, the actual occurrence of transport of ice crystals from the surface to the bottom of the Grand Banks has not been documented. However, an in-depth examination of past oceanographic and meteorological records might indicate the time periods when conditions most favorable to transport of ice to the sea bed over the Grand Banks occurred, and to what depth ice crystals might travel. One certain method by which ice crystals are carried to the bottom over the Grand Banks is by iceberg scouring of the sea bed.[77] This phenomenon is taken very seriously by the oil and gas industry, and is treated as a threat to equipment located at a depth of a 100 meters on the Grand Banks, Terra Nova Field. It seems likely that considerable quantities of ice crystals would be released into the water with the passage of a large, sea bed scouring iceberg, and any supercooled fish in the vicinity may be at risk of freezing.

Thus, while we cannot give examples of flounder loss due to ice, or survival due to the production of AFPs at depth on the Grand Banks, it is possible to make a case for the functionality of AFPs in the yellowtail flounder and American plaice that inhabit this area.

Physiological Ecology of AFGP — Future Directions

Fisheries Management — Identification of Physiologically Distinct Populations within the Northern Cod Stock

One of the major challenges faced by fisheries managers is to determine the boundaries within which populations of fish operate. These boundaries are sometimes not obvious. They may be rooted in environmental parameters, or differences in the biology or physiology of the populations under consideration. But, however they have evolved, they result in discrete stocks of fish which, while they may overlap in distribution, do not interbreed, and do not occupy identical environmental niches.

Traditionally, studies of tag returns, and morphometric and meristic data have been used to distinguish between populations.[53,54,78–80] In addition, biological/behavioral, genetic and physiological data are now being used to identify discrete populations and provide fisheries biologists with a more complete suite of tools with which to manage and conserve fish stocks.[69,81–87]

To date, population differences in Atlantic cod AFGP gene complement have not been investigated. However, plasma antifreeze levels have been used as a physiological time tag to distinguish between cod that have overwintered in subzero, inshore areas as opposed to those that have migrated inshore from warmer offshore overwintering sites. This information has, in turn, been coupled with analysis of microsatellite DNA, providing evidence of population structure at both Bank and Bay scales.[83,84] This work has helped support the contention that unique populations of Atlantic cod exist within the range that was previously considered as being occupied by a single stock complex for management purposes. Recently, Gilbert Bay, Labrador, the home of a distinct population of cod, has been put forward as an "Area of Intent", and is being considered for Marine Protected Area status under the Canada Oceans Act. As such, the discrete nature of the resident population will be recognized, and specific community based management plans will be developed for the area to avoid removal of these unique cod from the environment (J. Wroblewski, personal communication).

While Gilbert Bay appears to be a success story, the question must be asked, how many other discrete populations living at the edge of their distributional range, have been, or are in danger of being lost because their uniqueness has not been recognized? The Shinnecock Bay winter flounder may be genetically distinct from populations of winter flounder living along the adjacent open coastline, and a comparative study of such populations would prove informative from a variety of perspectives (evolutionary, management, conservation).

There is strong physiological evidence for the existence of Atlantic cod at the northern end of the distributional range with greater AFGP production capacity than any other population examined. Since the collapse of the cod fishery off the coast of Labrador and Newfoundland, both government and fishers have been waiting for signs of stock rebuilding. So far, in spite of optimistic predictions, this has not happened, and much reduced remnant populations living primarily in inshore locations along the northeastern coast of Newfoundland now appear to comprise the bulk of the Northern cod stock, a stock which once supported one of the major fisheries of the world. Even though numbers are vastly reduced, there is still pressure to resume fishing on what is left of the Northern cod stock. Under these circumstances, it becomes imperative to be armed with the best possible understanding of the reduced numbers of fish remaining. Are they all members of one interbreeding stock, or are there discrete units within the larger group? Mixed in with the fish currently residing inshore, is there a remnant group representing that part of the Northern cod stock that traditionally migrated offshore to overwinter? Would these fish be more likely to migrate offshore and re-establish their old migratory patterns, once density constraints were exceeded in the inshore area? Are any of the more cold-adapted northern individuals identified by Goddard *et al.*,[69] and Rose *et al.*[50] present in the inshore areas of Newfoundland? These questions can only be answered by further research. The numbers of Northern cod remaining are very low when compared to their previous status. For population rebuilding to take place successfully across the historical range, and to avoid the mistakes of the past, it is vital that all efforts be made to develop useful indicators of population structure, and then apply them to the remaining concentrations of cod in Newfoundland waters. Management strategies can then be developed based on whether the measures of population structure indicate homogeneity, or diversity within in the remaining concentrations of cod. Diversity in AFGP production capacity at the whole animal and at the genetic level would be one such measure of population structure. The development of molecular tools to allow examination Atlantic cod AFGP genes would be of great benefit in achieving a better understanding of cod stock structure, and to conserve genetic and physiological diversity of cold adapted populations.

This kind of approach should be applied to any species occurring over a wide distributional range that is being subjected to commercial fishing pressure. One such species that springs to mind is the lumpfish (*Cyclopterus lumpus)* which is fished for its roe. This species occurs in the Northwest Atlantic from Hudson Bay in the north, to Chesapeake Bay in the south. The adults undertake seasonal migrations, but the juveniles inhabit inshore shallow waters in the early years, a life cycle similar to the Atlantic cod. It is generally agreed that little is known about this species. In terms of population structure, biology, physiology, and possible AFP production in adults and juveniles, there is much to be discovered that would be invaluable to conservationists and fisheries managers when considering how to conduct any lumpfish fishery.

Response of Northern Species to Climate Change

The topic of climate change is controversial. Some areas of the globe are predicted to get warmer. However, others may become colder, and this is the prediction for the Northwest Atlantic region — an increase in the number of icebergs crossing the 48th parallel is factored into a report to the Canadian Environmental Assessment Agency considering development on the Terra Nova Field on the Grand Banks (Report of the Terra Nova Development Project Environmental Assessment Panel, August 1997). How might northern fish species be expected to react to such events?

The most rapid way in which fish can adjust to environmental change is by changing their distribution in relation to depth and temperature. Such changes might first be noticed by fishers and by research survey cruises. For example, in recent years, by-catch of American plaice in the Greenland halibut fishery has presented a problem to fisheries managers and conservationists. This is because some American plaice have moved into deeper water, and these two species are now co-existing on certain fishing grounds. The precise reason for this distributional shift is not known, but a protracted cold period in the region is thought to be a possible cause.[88]

As noted earlier, a snapshot of how antifreeze-producing fish might respond to changing climate can be seen in the case of the seven righteye flounder species living off the coast of Newfoundland. There appear to be three levels of freeze protection developed by these fish species — from no to little AFP production, to high levels produced early in the winter by the obligate producers inhabiting shallow waters.

AF(G)P production is generally matched to the environment inhabited. Will environmental shifts be mirrored by changes in fish population distributions, will current AF(G)P production systems in northern fish species be sufficient to allow survival in changing environments, or will we see changes in the patterns of AF(G)P production to allow survival and colonization of environments requiring additional freeze protection?

With the recent discovery of skin-type AFPs and their genes,[89,90] and with each new discovery seeming to pose new question, it is clear that the field of AF(G)P research is as dynamic, exciting, and changeable as the coastal ocean environment off Newfoundland and Labrador.

Acknowledgements

We thank JW Kiceniuk and J Ryder for the collection of biological samples from various fish species, and Paul Valerio and Fred Purton of the Huntsman Marine Science Centre for collection of the smooth flounder samples (data in Fig. 11, not previously published). This research was supported in part by grants to GLF from the Natural Sciences and Engineering Research Council of Canada (NSERC).

References

1. Holmes WN and Donaldson EM (1969). The body compartments and distribution of electrolytes. In: Hoar WS and Randall DJ (eds.), *Fish Physiology*, Vol. 1. Academic Press, New York, NY, pp. 1–98.
2. Prinsenberg SJ, Peterson IK, Narayanan S and Umoh JU (1997). Interaction between atmosphere, ice cover, and ocean off Labrador and Newfoundland from 1962 to 1992. *Can. J. Fish. Aquat. Sci.* **54**(Suppl. 1): 30–39.
3. Drinkwater KF, Pettipas R and Petrie L (1998). Overview of meteorological and sea ice conditions off Eastern Canada during 1997. *DFO Res. Doc.* 98/51.
4. Wroblewski JS, Collier A and Bailey W (1993). *Ocean temperatures in the Random Island region of Trinity Bay, 1991–1992.* The NSERC/Fishery Products International/National Sea Products Chair in Fisheries Oceanography, Data Rep. No. 1. Memorial University, St. John's, Nfld.: 48.
5. DeVries AL and Lin Y (1977). The role of glycoprotein antifreezes in the survival of Antarctic fishes. In: Llano GA (ed.), *Adaptations Within Antarctic Ecosystems: Proceedings of the 3rd Scientific Committee for Antarctic Research.* Smithsonian Institute, pp. 439–458.
6. Fletcher GL, King MJ and Kao MH (1987). Low temperature regulation of antifreeze glycopeptide levels in Atlantic cod (*Gadus morhua*). *Can. J. Zool.* **65**: 227–233.
7. Scholander PF, van Dam L, Kanwisher JW, Hammel HT and Gordon MS (1957). Supercooling and osmoregulation in Arctic fish. *J. Cell. Comp. Physiol.* **49**: 5–24.
8. Narayanan S, Carscadden J, Dempson JB, O'Connell MF, Prinsenberg S, Reddin DG and Shackell N (1995). Marine climate off Newfoundland and its influence on salmon (*Salmo salar*) and capelin (*Mallotus villosus*). In: Beamish RJ (ed.), Climate Change and Northern Fish Populations. *Can. Spec. Publ. Fish. Aquat. Sci.* No. 121: 461–474.

9. Reddin DG and Shearer WM (1987). Sea-surface temperature and distribution of Atlantic salmon in the Northwest Atlantic ocean. *Am. Fish. Soc. Symp.* **1**: 262–275.
10. Pinhorn AT (1976). Living marine resources of Newfoundland-Labrador: status and potential. *Bull. Fish. Res. Board Can.* **194**: 64.
11. Dempson JB and Krlstofferson AH (1987). Spatial and temporal aspects of the ocean migration of anadromous Arctic char. *Am. Fish. Soc. Symp.* **1**: 340–357.
12. Scott WB and Scott MG (1988). Atlantic fishes of Canada. *Can. Bull. Fish. Aquat. Sci.* **219**: 731.
13. Green JM and Farwell M (1971). Winter habits of the cunner, *Tautogolabrus adspersus* (Walbaum), in Newfoundland. *Can. J. Zool.* **49**: 1497–1499.
14. Valerio PF, Kao MH and Fletcher GL (1990). Thermal hysteresis activity in the skin of cunner, *Tautogolabrus adspersus. Can. J. Zool.* **68**: 1065–1067.
15. Green JM (1974). A localized mass winter kill of cunners in Newfoundland. *Can. Field Nat.* **88**: 1.
16. Fletcher GL (1975). The effects of capture "stress" and storage of whole blood on the red blood cells, total plasma proteins, glucose and electrolytes of the winter flounder (*Pseudopleuronectes americanus*). *Can. J. Zool.* **53**: 197–206.
17. Fletcher GL, Hew CL, Li X, Haya K and Kao MH (1985). Year-round presence of high levels of plasma antifreeze peptides in a temperate fish, ocean pout (*Macrozoarces americanus*). *Can. J. Zool.* **63**: 488–493.
18. Cheng CC and DeVries AL (1991). The role of antifreeze glycopeptides and peptides in the freezing avoidance of cold-water fish. In: di Prisco G (ed.), *Life Under Extreme Conditions.* Springer-Verlag, Berlin, Heidelberg, pp. 1–14.
19. Davies PL, Hew CL and Fletcher GL (1988). Fish antifreeze proteins: physiology and evolutionary biology. *Can. J. Zool.* **66**: 2611–2617.
20. DeVries AL (1983). Antifreeze peptides and glycopeptides in cold-water fishes. *Ann. Rev. Physiol.* **45**: 245–260.
21. Gordon MS, Amdur BN and Scholander PF (1962). Freezing resistance in some northern fishes. *Biol. Bull.* **122**: 52–62.
22. DeVries AL and Wohlschlag DE (1969). Freezing resistance in some Antarctic fishes. *Science* **163**: 1074–1075.

23. Rubinsky B, Arav A and Fletcher GL (1991). Hypothermic protection — a fundamental property of "antifreeze" proteins. *Biochem. Biophys. Res. Comm.* **180**: 566–571.
24. Negulescu PA, Rubinsky B, Fletcher GL and Machen TE (1992). Fish antifreeze proteins block Ca^{++} entry into rabbit parietal cells. *Am. J. Physiol.* **263**: C1310–1313.
25. Hochachka PW (1986). Defence strategies against hypoxia and hypothermia. *Science* **231**: 23–241.
26. Hochachka PW (1988). Metabolic-, channel-, and pump-coupled functions: constraints and compromises of coadaptation. *Can. J. Zool.* **66**: 1015–1027.
27. Hochachka PW (1988). Channels and pumps — determinants of metabolic cold adaptation strategies. *Comp. Biochem. Physiol.* **90B**: 515–519.
28. Hays LM, Feeney RE, Crowe LM, Crowe JH and Oliver AE (1996). Antifreeze glycoproteins inhibit leakage from liposomes during thermotropic phase transitions. *Proc. Natl. Acad. Sci. USA* **93**: 6835–6840.
29. Wu Y and Fletcher GL (2000). Efficacy of antifreeze protein types in protecting liposome membrane integrity depends on phospholipid class. *Biochem. Biophys. Acta (General Subjects)* **1524**: 11–16.
30. Wu Y, Banoub J, Goddard SV, Kao MH and Fletcher GL (2001). Antifreeze glycoproteins: relationship between molecular weight, thermal hysteresis and the inhibition of leakage from liposomes during thermotropic phase transition. *Comp. Biochem. Physiol. B* (in press).
31. Tablin F, Oliver AE, Walker NJ, Crowe LM and Crowe JH (1996). Membrane phase transition of intact human platelets: correlation with cold-induced activation. *J. Cell. Physiol.* **168**: 305–313.
32. Hobbs KD and Fletcher GL (1996). Antifreeze proteins increase the cold tolerance of goldfish. *Bull. Can. Soc. Zool.* **27**: 64.
33. Hobbs KD (1999). *The Effect of Antifreeze Proteins on the Cold Tolerance of Goldfish* (*Carassius auratus* L.) M.Sc. Thesis, Department of Biology, Memorial University of Newfoundland: 70.
34. Fletcher GL, Haya K, King MJ and Reisman HM (1985). Annual antifreeze cycles in Newfoundland, New Brunswick and Long Island winter flounder, *Pseudopleuronectes americanus. Mar. Ecol. Prog. Ser.* **21**: 205–212.
35. DeVries AL (1971). Glycoproteins as biological antifreeze agents in Antarctic fishes. *Science* **172**: 115–1155.

36. DeVries AL (1988). The role of antifreeze glycopeptides and peptides in the freezing avoidance of Antarctic fishes. *Comp. Biochem. Physiol.* **90B**: 611–621.
37. Kao MH, Fletcher GL, Wang NC and Hew CL (1986). The relationship between molecular weight and antifreeze polypeptide activity in marine fish. *Can. J. Zool.* **64**: 578–582.
38. Fletcher GL (1981). Effects of temperature and photoperiod on the plasma freezing point depression, Cl^- concentration, and protein "antifreeze" in winter flounder. *Can. J. Zool.* **59**: 193–201.
39. Fourney RM, Fletcher GL and Hew CL (1984). The effects of long daylength on liver antifreeze MRNA in the winter flounder (*Pseudopleuronectes americanus*). *Can. J. Zool.* **62**: 1456–1460.
40. Fletcher GL (1977). Circannual cycles of blood plasma freezing point and Na^+ and Cl^- concentrations in Newfoundland winter flounder (*Pseudopleuronectes americanus*): correlation with water temperature and photoperiod. *Can. J. Zool.* **55**: 789–795.
41. Idler DR, Fletcher GL, Belkhode S, King MJ and Hwang SJ (1989). Regulation of antifreeze protein production in winter flounder: a unique function for growth hormone. *Gen. Comp. Endocrinol.* **74**: 327–334.
42. Davies PL, Fletcher GL and Hew CL (1999). Freeze-resistance strategies based on antifreeze proteins. In: Storey KB (ed.), *Environmental Stress and Gene Regulation*, BIOS Scientific Publishers Ltd, pp. 61–80.
43. Fletcher GL and Smith JC (1980). Evidence for permanent population differences in the annual cycle of plasma "antifreeze" levels of winter flounder. *Can. J. Zool.* **58**: 507–512.
44. Hew CL, Slaughter D, Fletcher GL and Joshi SK (1981). Antifreeze glycoproteins in the plasma of Newfoundland Atlantic cod (*Gadus morhua*). *Can. J. Zool.* **59**: 2186–2192.
45. Goddard SV, Wroblewski JS, Taggart CT, Howse KA, Bailey WL, Kao MH and Fletcher GL (1994). Overwintering of adult northern Atlantic cod (*Gadus morhua*) in cold inshore waters as evidenced by plasma antifreeze glycoprotein levels. *Can. J. Fish. Aquat. Sci.* **51**: 2834–2842.
46. Goddard SV and Fletcher GL (1994). Antifreeze proteins: their role in cod survival and distribution from egg to adult. *ICES Mar. Sci. Symp.* **198**: 676–683.

47. Kao MH and Fletcher GL (1988). Juvenile Atlantic cod (*Gadus morhua*) can be more freeze resistant than adults. *Can. J. Fish. Aquat. Sci.* **45**: 902–905.
48. Goddard SV, Kao MH and Fletcher GL (1992). Antifreeze production, freeze resistance, and overwintering of juvenile Northern Atlantic cod (*Gadus morhua). Can. J. Fish. Aquat. Sci.* **49**: 516–522.
49. Goddard SV, Morgan, MJ and Fletcher GL (1997). Influence of plasma antifreeze glycoproteins on temperature selection by Atlantic cod (*Gadus morhua*) in a thermal gradient. *Can. J. Fish. Aquat. Sci.* **54**(Suppl. 1): 88–93.
50. Rose GA, deYoung B, Kulka DW, Goddard SV and Fletcher GL (2000). Distributional shifts and overfishing the Northern cod (*Gadus morhua*): a view from the ocean. *Can. J. Fish. Aquat. Sci.* **57**: 644–664.
51. Lear HW and Green JM (1984). Migration of the Northern Atlantic cod and the mechanisms involved. In: McCleave JD, Arnold GP, Dodson JJ and Neill WH (eds.), *Mechanisms of Migration in Fishes.* Plenum Publishing Corporation, pp. 309–315.
52. Rose GA (1993). Cod spawning on a migration highway in the northwest Atlantic. *Nature* **366**: 458–461.
53. Templeman W (1974). Migration and intermingling of Atlantic Cod (*Gadus morhua*) Stocks of the Newfoundland area. *J. Fish. Res. Bd. Can.* **31**(6): 1073–1092.
54. Templeman W (1979). Migration and intermingling of stocks of Atlantic cod *Gadus morhua*, of the Newfoundland and adjacent areas from tagging in 1962–66. *Int. Comm. Northwest Atl. Fish. Res. Bull.* No. 14: 50.
55. Taggart CT and Frank KT (1987). Coastal upwelling and *Oikopleura* occurrence ("Slub"): a model and potential application to inshore fisheries. *Can. J. Fish. Aquat. Sci.* **44**: 1729–1738.
56. Clark DS and Green JM (1990). Activity and movement patterns of juvenile Atlantic cod, *Gadus morhua*, in Conception Bay, Newfoundland, as determined by sonic telemetry. *Can. J. Zool.* **68**: 143–1442.
57. Dalley EL and Anderson JT (1997). Age-dependent distribution of demersal juvenile Atlantic cod (*Gadus morhua*) in inshore/offshore northeast Newfoundland. *Can. J. Fish. Aquat. Sci.* **54**(Suppl. 1): 168–176.
58. Brown JA, Pein P, Methvan DA and Somerton DC (1989). The feeding, growth, and behaviour of juvenile cod, *Gadus morhua* L., in cold environments. *J. Fish. Biol.* **35**: 373–380.

59. Bogstad B, Lilly GR, Mehl S, Pálsson OK and Stefánsson G (1994). Cannibalism and year-class strength in Atlantic cod (*Gadus morhua* L.), in Arcto-boreal ecosystems (Barents Sea, Iceland, and eastern Newfoundland). *ICES Mar. Sci. Symp.* **198**: 576–599.
60. Chadwick EMP, Cairns DK, Dupuis HMC, Ewart KV, Kao MH and Fletcher GL (1990). Antifreeze levels reflect the migratory behaviour of Atlantic Herring (*Clupea harengus harengus*) in the Southern Gulf of St. Lawrence. *Can. J. Fish. Aquat. Sci.* **47**: 1534–1536.
61. Fletcher GL, Goddard SV, Davies PL, Gong Z, Ewart KV and Hew CL (1998). New insights into fish antifreeze proteins: physiological significance and molecular regulation. In: Portner HO and Playle R (eds.), *Cold Ocean Physiology* (Society for Experimental Biology seminar series: 65). Cambridge University Press, Cambridge, UK, pp. 239–265.
62. Fletcher GL, Hew CL and Davies PL (2001). Antifreeze proteins of teleost fishes. *Ann. Rev. Physiol.* **63**: 359–390.
63. Hayes PH, Davies PL and Fletcher GL (1991). Population differences in antifreeze protein gene copy number and arrangement in winter flounder. *Genome* **34**: 174–177.
64. Scott GK, Hew CL and Davies PL (1985). Antifreeze protein genes are tandemly linked and clustered in the genome of the winter flounder. *Proc. Natl. Acad. Sci. USA* **82**: 2613–2617.
65. Fletcher GL, Kao MH and Fourney RM (1986). Antifreeze peptides confer freezing resistance to fish. *Can. J. Zool.* **64**: 189–1901.
66. Fletcher GL, Kao MH and Haya K (1984). Seasonal and phenotypic variations in plasma protein antifreeze levels in a population of marine fish, sea raven (*Hemitripterus americanus*). *Can. J. Fish. Aquat. Sci.* **41**: 819–824.
67. Hew CL, Wang NC, Joshi S, Fletcher GL, Scott GK, Hayes PH, Buettner B and Davies PL (1988). Multiple genes provide the basis for antifreeze protein diversity and dosage in the ocean pout, *Macrozoarces americanus. J. Biol. Chem.* **263**: 12049–12055.
68. Leim AH and Scott WB (1966). Fishes of the Atlantic Coast of Canada. *Fish. Res. Bd. Can. Bull.* No. 155: 485.
69. Goddard SV, Kao MH and Fletcher GL (1999). Population differences in antifreeze production cycles of juvenile Atlantic cod (*Gadus morhua*) reflect adaptations to overwintering environment. *Can. J. Fish. Aquat. Sci.* **56**: 1991–1999.

70. Taggart CT, Anderson J, Bishop C, Colbourne E, Hutchings J, Lilly G, Morgan J, Murphy E, Myers R, Rose G and Shelton P (1994). Overview of cod stocks, biology, and environment in the Northwest Atlantic region of Newfoundland, with emphasis on northern cod. *ICES Mar. Sci. Symp.* **198**: 140–157.
71. Davies PL, Hough C, Scott GK, Ng N, White BN and Hew CL (1984). Antifreeze protein genes of the winter flounder. *J. Biol. Chem.* **259**: 924–9247.
72. Scott GK, Fletcher GL and Davies PL (1986). Fish antifreeze proteins: recent gene evolution. *Can. J. Fish. Aquat. Sci.* **43**: 1028–1034.
73. Scott GK, Davies PL, Kao MH and Fletcher GL (1988). Differential amplification of antifreeze protein genes in the pleuronectinae. *J. Mol. Evol.* **27**: 29–35.
74. Drinkwater KF (1994). Overview of environmental conditions in the Northwest Atlantic in 1993. *NAFO SCR Doc.* 94/20: 61.
75. Drinkwater KF, Petrie B and Narayanan S (1992). Overview of environmental conditions in the Northwest Atlantic in 1991. *NAFO SCR Doc.* 92/73: 36.
76. Templeman W (1972). Temperatures and salinities in the eastern Newfoundland area in 1971. *ICNAF Res. Doc.* 72/31: 19–25.
77. Davidson S, Simms A, Sönnichsen G and Clark P (1997). Characterization of iceberg pits on the Grand Banks of Newfoundland. In: *Proceedings of Oceans '97, Halifax, Nova Scotia, CA, 6–9 Oct 1997,* Vol. 2. MTS/IEEE, p. 1394.
78. Myers RA, Barrowman NJ and Hutchings JA (1997). Inshore exploitation of Newfoundland Atlantic cod (*Gadus morhua*) since 1948 as estimated from mark-recapture data. *Can. J. Fish. Aquat. Sci.* **54**(Suppl. 1): 224–235.
79. Taggart CT, Penny P, Barrowman N and George C (1995). The 1954–1993 Newfoundland cod-tagging database: statistical summaries and spatial-temporal distributions. *Can. Tech. Rep. Fish. Aquat. Sci.* No. 2042.
80. Templeman W (1981). Vertebral numbers in Atlantic cod, *Gadus morhua*, of the Newfoundland and adjacent areas, 1947–71 and their use for delineating cod stocks. *J. Northwest Atl. Fish. Sci.* **2**: 21–45.
81. Bentzen P, Taggart CT, Ruzzante D and Cook D (1996). Microsatellite polymorphism and the population structure of Atlantic cod (*Gadus morhua*)

in the northwest Atlantic. *Can. J. Fish. Aquat. Sci. DFO Atl. Fish. Res. Doc.* 96/44: 24

82. Campana SE, Chouinard GA, Hanson JM, Fréchet A and Brattey J (2000). Otolith elemental fingerprints as biological tracers of fish stocks. *Fish. Res.* **46**: 343–357.
83. Ruzzante DE, Taggart CT, Cook D and Goddard SV (1996). Genetic differentiation between inshore and offshore Atlantic cod (*Gadus morhua* L.) off Newfoundland: microsatellite DNA variation and antifreeze level. *Can. J. Fish. Aquat. Sci.* **53**: 634–645.
84. Ruzzante DE, Wroblewski JS, Taggart CT, Smedbol RK, Cook D and Goddard SV (2000). Bay-scale population structure in coastal Atlantic cod in Labrador and Newfoundland, Canada. *J. Fish. Biol.* **56**: 431–447.
85. Shaklee JB and Bentzen P (1998). Genetic identification of stocks of marine fish and shellfish. *Bull. Mar. Sci.* **62**: 589–621.
86. Smedbol RK and Wroblewski JS (1997). Evidence for inshore spawning of northern Atlantic cod (*Gadus morhua*) in Trinity Bay, Newfoundland, 1991–1993. *Can. J. Fish. Aquat. Sci.* **54**(Suppl. 1): 177–186.
87. Taggart CT, Ruzzante DE and Cook D (1998). Localized stocks of cod (*Gadus morhua*) in the Northwest Atlantic: the genetic evidence and otherwise. In: Hunt von Herbing I, Kornfield I, Tupper M and Wilson J (eds.), *The Implications of Localized Fishery Stocks*, National Resource, Agriculture, and Engineering Service, Ithaca, New York, pp. 65–90.
88. Avila de Melo AM and Alpoim R (1999). Northwest Atlantic fisheries resources: analysis of the present situation based on the NAFO Scientific Council 1998 evaluations. *Relat. Cient. Tec. Inst. Pescas Mar.* No. 35: 24.
89. Gong Z, Ewart KV, Hu Z, Fletcher GL and Hew CL (1996). Skin antifreeze protein genes of the winter flounder *Pleuronectes americanus*, encode distinct and active polypeptides without the secretory signal and prosequence. *J. Biol. Chem.* **271**: 4106–4112.
90. Gong Z, King, MJ, Fletcher GL and Hew CL (1995). The antifreeze protein genes of the winter flounder, *Pleuronectes americanus*, are differentially regulated in liver and non-liver tissues. *Biochem. Biophys. Res. Comm.* **206**: 387–392.

Chapter 3

Fish Antifreeze Proteins: Functions, Molecular Interactions and Biological Roles

Kathryn Vanya Ewart
NRC Institute for Marine Biosciences
1411 Oxford St., Halifax, NS, B3H 3Z1, Canada

Introduction

Antifreeze proteins (AFPs) and antifreeze glycoproteins (AFGPs) bind to ice crystals and inhibit crystal growth. AF(G)Ps lower the melting point of a solution colligatively, as any solute would, according to their concentration. However, because of the interaction of AF(G)Ps with ice crystals, they can generate a very efficient non-colligative freezing point depression that far exceeds the depression predicted based on colligative properties alone. This results in a difference between the freezing and melting points that is called a thermal hysteresis. AF(G)Ps can also inhibit ice recrystallization, which refers to the coalescence of small ice crystals into larger ones at variable subzero temperatures. Although the mechanism of action of the AF(G)Ps is not precisely understood at the molecular level, it appears that their effects in solutions all arise from their direct interaction with ice crystals.

A number of different AF(G)Ps have been identified in bacteria, plants, fungi and animals.[1] However, the greatest diversity of AF(G)Ps found to date is in teleost fish and the fish AF(G)Ps have also been the best studied. AF(G)Ps are widely distributed among fish species that inhabit polar and north temperate oceans and offer protection from freezing in icy seawater. Teleost fishes are at risk of freezing in cold oceans because they have a lower solute concentration than the surrounding seawater. Solute molecules lower solution freezing points colligatively according to their concentration in solution. As a result of their total dissolved solutes, fishes have freezing points generally ranging from –0.6 to –0.8°C. However, seawater can have a freezing point as low as –2°C, depending on salinity. Because of this, when sea water reaches its equilibrium freezing point, fish that live in the water are below their freezing points. If ice is present, it can propagate across the fish, which would then freeze. Species such as righteye flounders, herrings, smelts, eelpouts, gadids and others have biochemical adaptations that allow them to inhabit the icy waters without the risk of freezing. The non-colligative AF(G)Ps are the most widespread and diverse of these adaptations and these proteins can lower the freezing points of fish to that of the seawater (Fig. 1).

Studies of AF(G)Ps and of the fish species that produce them are complementary. Investigations of the fish AF(G)Ps have contributed to our understanding of the freeze resistance adaptations, overwintering strategies and life histories of various fish species. For species such as Atlantic cod (*Gadus morhua*), AF(G)P research has provided valuable information for aquaculture and fisheries applications.[2,3] At the same time, investigation of the different AF(G)Ps (and their genes) from several fish species has revealed intriguing new insights into the biochemistry and mechanism of action of the AF(G)Ps and this

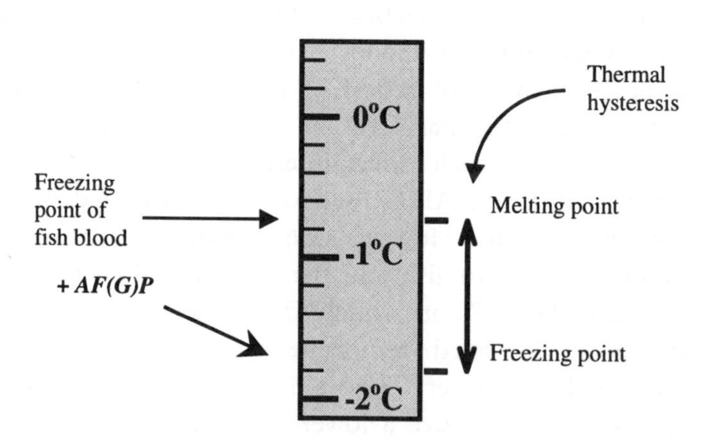

Fig. 1 Diagram outlining the non-colligative effect of AF(G)P on freezing points in fish.

knowledge will be valuable in the development of biotechnology applications for the AF(G)Ps. Some of the salient findings on AF(G)Ps are brought together here in order to review the actions of AF(G)Ps — from their molecular functions to their roles in fish.

Diversity and Evolution of the AF(G)Ps

Five structural types of AF(G)Ps have been identified to date in fish and there is considerable diversity even within each type. These proteins types include AFPs I to IV and the antifreeze glycoproteins (AFGPs). The classification of these proteins and their characteristics are outlined in Table 1. The proteins are remarkably diverse and each type is

Table 1 Diversity of the fish AF(G)Ps.

Protein	Species	Characteristics	Similarity to other proteins	References
Type I	Sculpins, righteye flounders	α-helical, some with repeat motif	None found	6–10
Type II	Herring, smelt, sea raven	Globular, first two are Ca^{2+}-dependent	C-type lectins	11–13
Type III	Eel pouts	Globular with one flat surface	None found	14–16
Type IV	Longhorn sculpin	Folded α-helical structure	Low density lipoprotein receptor-binding domain of apolipoprotein E	17
AFGP	Cods, Antarctic notothenioids	Ala-Ala-Thr polymer with disaccharide on each Thr	None found*	18

*The gene for the notothenioid AFGP is, however, related to trypsinogen (see text).

narrowly distributed among fish groups. The unusual combination of structural diversity and limited distribution among fish AF(G)P types led to a hypothesis suggesting that each type (and possibly some AF(G)Ps within the five types) had evolved independently, well after the taxonomic groups of fish has attained their current forms.[4] It was proposed that over the course of geologically recent cooling and glaciation events, fish species would have undergone strong selective pressure to avoid freezing.[4] Selection would have favored any mechanism of lowering the freezing point that would not disrupt homeostasis and this could have led to the variety of AF(G)Ps that exist in fish groups now studied.

The different fish AF(G)Ps appear to have two contrasting origins. Some have emerged as ice-binding variants in protein families known to have unrelated functions. Others are genuinely new proteins whose genes came into existence through recent genetic events. The antifreeze glycoproteins of Antarctic notothenioids are an example of the latter "new" proteins and are discussed elsewhere in this volume. These glycoprotein genes evolved by repeated duplication of a nine-base pair segment of DNA at an intron/exon boundary in a trypsinogen gene[5] and the encoded glycoproteins bear no similarity to trypsinogen.

The emergence of AF(G)Ps as ice-binding variants in other protein families is exemplified by the type II AFPs. Three fish AFPs, those of rainbow smelt (*Osmerus mordax*), herring (*Clupea harengus harengus*) and sea raven (*Hemitripterus americanus*) are classified as type II and they are all homologous to the carbohydrate-recognition domains (CRDs) of C-type lectins (Table I). Models of the AFPs based on the crystal structures of the rat mannose-binding protein and E-selectin show strong similarity between the folds of the lectins and those of the AFPs.[19,20] Moreover, the solution structure of sea raven AFP resolved by nuclear magnetic resonance further confirms this similarity.[21] The C-type animal lectins comprise a large protein superfamily. C-type CRDs are found in a variety of proteins with structural, immunological and metabolic roles.[22] The soluble C-type lectins of vertebrates include mannose-binding lectins and other collectins that play roles in innate immunity involving recognition of pathogen by their surface carbohydrates.[23] Soluble galactose-binding C-type lectins are thought to play a similar role in invertebrates.[24] It is not yet known which of these lectins gave rise to the type II AFPs. However, recent findings on the different type II AFPs suggest that antifreeze activity evolved more than once, through minor changes in separate C-type lectins. The AFPs of smelt and herring are approximately 80% identical. Both of these proteins bind Ca^{2+} and require it for antifreeze activity.[11,12] In these two AFPs, the precise residues corresponding to the galactose-specific C-type CRD Ca^{2+} and carbohydrate-binding site are conserved.[11,12] Moreover, the ice-binding site of the herring AFP was shown to correspond to the galactose-binding site of corresponding C-type lectins.[25] It would appear that a C-type lectin/AFP with a biological function different from ice-binding might have existed in smelt and herring and then evolved an AFP function. In contrast, the AFP of sea raven is only approximately 40% identical to the herring or smelt AFPs.[12] The galactose-binding Ca^{2+} and carbohydrate-binding site of C-type lectins is not conserved in the sea raven[12] and the ice-binding site of this protein is located on a different surface of the domain from that of the herring AFP.[26] Thus, the Ca^{2+}-dependent type II AFPs of smelt and herring AFPs appear to have evolved separately from the Ca^{2+}-independent AFP of sea raven.[1] These proteins all seem to have emerged through minor changes in distinct CRDs that led to different ice-binding sites.

Other AF(G)Ps such as the alpha-helical type I AFPs or the small globular type III AFPs have no known homologues other than related AFPs. The evolutionary progenitors of these proteins remain to be discovered. One of the difficulties in searching for proteins related to the type I AFPs is the preponderance of Ala residues in their sequences. Bioinformatics software designed to search databases for homologous sequences can give spurious results as a result of the unusual Ala frequency in these proteins, either failing to identify genuine matches when the weight given to Ala is too low or generating false matches because of excessive Ala overlap. It is also possible that no homologues exist for type I AFPs. As the type I serum AFP of flounder are composed of repeating unit (Thr-X_{10}, with most X being Ala), they may well have evolved *de novo* as the notothenioid AFGPs did. This question is as yet unresolved.

In the midst of this myriad of proteins and origins, the unifying factors are (1) that all these proteins appear to serve the same general purpose in fish, and (2) that they all appear to function in the same manner.

Effects of AF(G)Ps and Their Measurement

The binding of AF(G)Ps to ice surfaces results in thermal hysteresis, as noted above, as well as an inhibition of ice recrystallization. Because AF(G)Ps bind to ice in a geometrically defined fashion, they stop crystal growth in specific directions and thereby alter ice morphology. The change in morphology of ice crystals caused by an AF(G)P is shown in Fig. 2. All AF(G)Ps have similar effects on ice crystal morphology but the specific morphologies generated differ among the antifreeze types. The activity of AF(G)Ps is most commonly detected and measured as thermal hysteresis. This measurement is made by generating a solution with a seed ice crystal and determining the melting and freezing points as detected by ice growth and ice melting. The apparatus most frequently used is the Clifton nanolitre osmometer (Clifton Technical Physics, Hartford NY). It consists of a cooling stage mounted for viewing on a microscope stage. The stage has a sample holder that contains wells that can take very small (<1 μl) samples for observation. Sample temperature is regulated by a Peltier cooling device that is under very fine manual control. Although the cooling stage is meant to be viewed under a stereomicroscope, most researchers mount it on a compound microscope in order to obtain sufficient magnification to observe ice crystal morphologies and record images. Samples are frozen and then thawed until a single crystal is evident. The crystals are normally between 25 and 100 μm in length in this system. This is much larger than the incipient crystals that would be expected in sea water and in fish. However, the *in vitro* growth measurements made on these larger crystals do appear to be relevant to the *in vivo* situation. Thermal hysteresis can be measured by recording the melting and freezing points

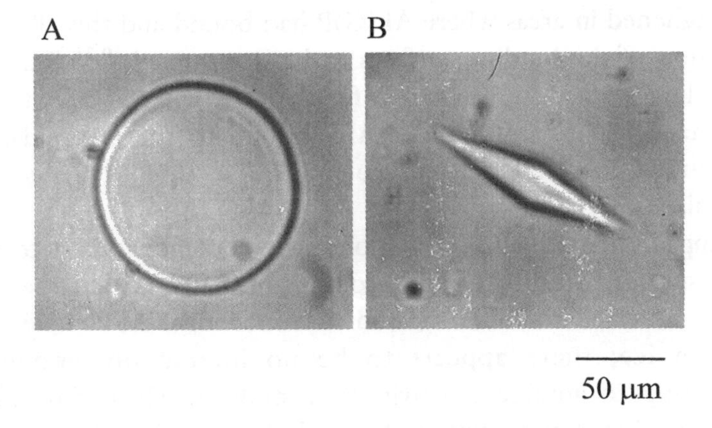

Fig. 2 Ice crystal morphology in the absence (**A**) and presence (**B**) of AF(G)P as viewed using the nanolitre osmometer and a compound microscope. In (A), the C axis is perpendicular to the page whereas in (B) it is parallel to the page and is the long axis of the crystal. The protein used was rainbow smelt AFP.

observed as ice crystal growth and melting.[27] At temperatures in the hysteresis interval, ice crystal morphologies can also be viewed and recorded digitally. The prism planes become evident on ice crystals in solutions containing AF(G)Ps at concentrations two to three orders of magnitude lower than those at which full hysteresis is evident.[28,29] The advantages of measuring thermal hysteresis are that it is a quantitative measurement and its magnitude also corresponds to meaningful freeze protection in fish. However, the analysis of ice crystal morphology is far more sensitive and, as will be discussed below, has become very informative.

Activity of AF(G)Ps at the Ice Surface

The fish AF(G)Ps share the same general effect on ice, causing changes in morphology and thermal hysteresis. However, at the molecular level, the AF(G)Ps differ in binding to distinct and specific surfaces of ice crystals. This was demonstrated in a series of experiments in which ice hemispheres were grown in dilute AF(G)P solutions. The ice surface was roughened in areas where AF(G)P had bound and this allowed the orientation of the binding surfaces to be determined.[30,31] Remarkably, the binding surfaces were distinct for the different AF(G)Ps tested. In one instance, a winter flounder AFP bound to the pyramidal planes while, in contrast, the related sculpin AFP bound to the secondary prism planes.[31]

It appears that the AF(G)P molecules bind individually to the ice crystal surface. Studies using recombinant AFPs with large protein domains attached to them showed that individual AFP molecules act alone on ice; there appears to be no interaction among AFPs contributing to antifreeze activity.[32] A study in which different AFP types were mixed and activity observed also supports this conclusion. Results using type I AFP from winter flounder (*Pleuronectes americanus*), type II from sea raven and type III from ocean pout (*Macrozoarces americanus*) showed that the ice morphologies generated with mixtures were intermediate between the types used and that the antifreeze activity was additive.[32,33] In addition, the first direct molecular-level investigation of AF(G)P action at the ice surface showed a pattern consistent with individual AF(G)P action. The surface of ice in a solution of winter flounder type I AFP was probed by scanning tunneling microscopy (STM) and compared with an equivalent ice surface devoid of AFP. Grooves were evident on the surface image obtained in the presence of AFP (Fig. 3).[34] The grooves were aligned approximately 65° to one hexagonal side on the pyramidal plane, consistent with the known orientation of flounder AFP on ice.[34] The dimensions of the grooves were 6 ± 1 nm in length and 2 ± 1 nm in width, which correspond to the size of a single AFP molecule (50 × 10 Å).[34,35] The sizes and orientations of the AFP-associated grooves are relatively clear and there appears to be no association between them.

Studies on ice structure have shown that no sharp interface exists between ice and water that would be defined as an ice crystal surface but, rather, a gradual transition layer is observed between liquid water and solid ice [reviewed in Petren Ko (1994)[36]]. In this context, it will be important to determine the level in the transition layer at which the AFP settles and what effect, if any, interaction of ice with AFPs might

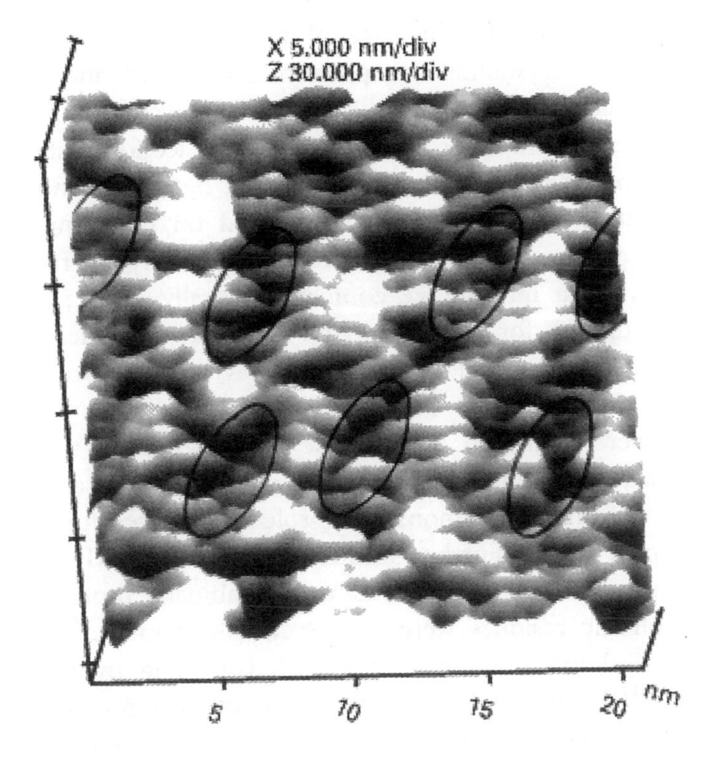

Fig. 3 STM surface plot of ice in the presence of type I AFP. The grooves that correspond to AFP dimensions are indicated with circles. Reprinted with permission from Grandum *et al.* (1999).[34]

have on this layer. More thorough STM studies may provide insight into this, particularly as the STM study suggests that the positions of AFP molecules can be recognized. However, it will be important to determine the nature of the apparent surface detected by STM before any conclusions can be drawn using this approach.

Biochemical studies on AF(G)P are generating another paradigm shift in our understanding of AFP-ice interaction. In spite of extensive work on the structure and function of AFPs including the determination of 3D structures for three of the fish AFP types and extensive functional studies of most of them, it is still not clear what precisely the

AFP-ice contacts are. In the case of many binding proteins, it is customary to co-crystallize the protein and ligand and determine molecular contacts from the crystal structure of the complex. Because it is impossible to prepare stable crystals of AFP and ice for cyrstallography, these direct means of determining the precise ligands are not feasible. As an alternative, many studies have focused on preparing synthetic analogues or mutants of the AFPs in order to identify the residues and specific groups that mediate ice contact. Generally, these studies have been successful and, for types I and III AFPs, key residues have been identified. However, much of this work was based on the incorrect assumption that the sole or predominant mediators of AFP-ice interaction were hydrogen bonds. The first suggestion that van der Waals interactions might contribute to the activity of an antifreeze protein arose from the demonstrated role of Leu residues in winter flounder type I AFP.[37] Further studies on types I and III AFPs using both synthetic analogues and mutant recombinant proteins revealed that hydrophobic residues were directly involved in ice-binding.[38–40] These empirical findings are in line with reasoning by Haymet and colleagues.[41,42] Earlier models for protein-ice interaction considered only hydrogen bonding and only accounted for the favorable enthalpy of interaction between the protein and ice. These models neglected to consider that, when not bound to ice, the same groups could hydrogen bond to liquid water molecules.[39] Moreover, it is likely that hydrogen bonding is actually favored in water over ice because the spatial constraints in ice would be restrictive.[42] Considering these arguments, it is clear that hydrogen bonding alone would not be likely to mediate strong ice-binding by the AF(G)Ps. The interaction with ice would be more likely to be at least partially hydrophobic. Thus, the experimental results showing roles for hydrophobic residues can be explained in this context. This new paradigm is leading to greater insight into the structure and function of AF(G)Ps by leading researchers to ask different questions. For example, in an elegant study, the roles of hydrophobic residues in the winter flounder type I AFP were investigated.[43] This study revealed that the helix surface that actually binds to ice is not the one that researchers had been focusing on in functional studies over the past two decades. This discovery will bring us closer to an understanding of precisely how AF(G)Ps and ice interact.

Action of AF(G)Ps in Fish

As the molecular level functioning of AF(G)Ps is deciphered, our understanding of the larger question of how and where AF(G)Ps act in the fish is also progressing. Physiological, molecular and immunological studies have provided insight into the role of AFP in freeze resistance and how ice growth is prevented at the whole-animal level.

The location of AF(G)Ps within the fish can be informative in determining their mechanism of action. AF(G)Ps are normally purified from blood plasma or serum for ease of collection but they are also present in other tissues. Investigation of an Antarctic Notothenioid species showed that the AFGP was present in all tissues except brain.[44] The smallest AFGP isoforms were also secreted into the intestinal fluids.[45] Notothenioids, which have an aglomerular kidney, appear to retain their AFGP whereas AFPs are excreted into the urine of cod, ocean pout, sea raven and winter flounder.[46] AFPs were also detected in the skin of winter flounder and shorthorn sculpin (*Myoxocephalus scorpius*).[47,48] AFP levels in flounder skin were found to be approximately equal to those in blood.[47] Initially, all AF(G)Ps appeared to be extracellular. Moreover, the cDNAs isolated for these proteins all revealed signal sequences which are protein tags for secretion from the cell.[49] However, as discussed elsewhere in this volume, the AFP cDNAs recently cloned from the skin of flounder and sculpin revealed no signal sequence for secretion from the cell.[50,51] This suggested that the encoded AFPs are intracellular.

There are two alternative hypotheses for the role of AF(G)Ps in preventing ice growth in fish and different findings on the location of AF(G)Ps can be used to support either of them. The ice-exclusion hypothesis maintains that AF(G)P in skin, gills and other external surfaces as well as in the digestive system form a barrier beyond which ice crystals cannot grow and that ice crystals in the surrounding water would therefore never get into the fish. The tissues and fluids within the fish would thereby be kept ice-free. The presence of high levels of AFP in the skin of winter flounder would support this ice exclusion mechanism. In addition, this mechanism would offer an explanation for the occurrence of intracellular AFPs in skin. The intracellular and extracellular AFPs in skin might together ensure a continuous AFP distribution. Moreover, the skin itself provides a physical barrier to ice growth.[47] However, even if the skin of fish can serve as a formidable barrier to ice crystal growth, it cannot offer complete protection from freezing. Ice may still form within the fish and ice from the surrounding water may enter the fish through the gut, the gill or skin wounds and abrasions.[47,50]

The alternative hypothesis for antifreeze action is that AF(G)Ps act within the fish body to bind small ice crystals and prevent their growth. According to this hypothesis, tiny ice crystals may be present within the fish but AF(G)Ps would bind to the crystals and stop their growth, thereby preventing freezing of the animal. This mode of AF(G)P action is consistent with the high levels of AF(G)P circulating in the blood of fish. This hypothesis was tested by determining whether there is evidence for ice within fish. Cheng and DeVries have obtained evidence for the presence of ice crystals within fish through whole-animal freezing experiments.[5] In this work, fish were transferred to water that was ice-free and then they were supercooled (i.e. brought to a temperature below their freezing point). When the water temperature was lowered beyond the maximum AFGP-induced freezing point depression, the fish rapidly froze. This freezing could only take place if ice was present within the fish.[5] Indirect, but equally persuasive, evidence for the presence of ice within fish has come from immunological work modeled on the previous discovery by Addadi and colleagues of immune responses to urate and other small organic crystals leading to the production of anti-crystal immunoglobulins (Igs).[52] Based on these studies, it seemed reasonable to postulate that if ice crystals are present in fish, these crystals should generate an immune response and anti-ice Igs should be produced as a result. To test for the presence of anti-ice Igs, the strategy used for urate was adopted.[52] If an Ig recognizes a crystal, then it should have a binding surface complementary to the crystal lattice. As a result, the Ig should form a template upon which the crystal can grow readily and it should therefore foster crystal nucleation. To determine if this was the case, ice crystal nucleation assays were performed. The samples included sera from four ice-free species including a rat, freshwater rainbow trout (*Oncorhynchus mykiss*), bighead carp (*Aristichthys nobilis*) and tilapia (*Oreochromis niloticus*) and two freeze-resistant North Atlantic marine species known to be exposed to icy seawater (Atlantic herring and ocean pout). Ice nucleation activity was detected in the sera of the North Atlantic species but not the others (Fig. 4).[53] Pure Ig fractions obtained from ocean pout serum by two different methods also showed ice nucleation activity.[53] These results

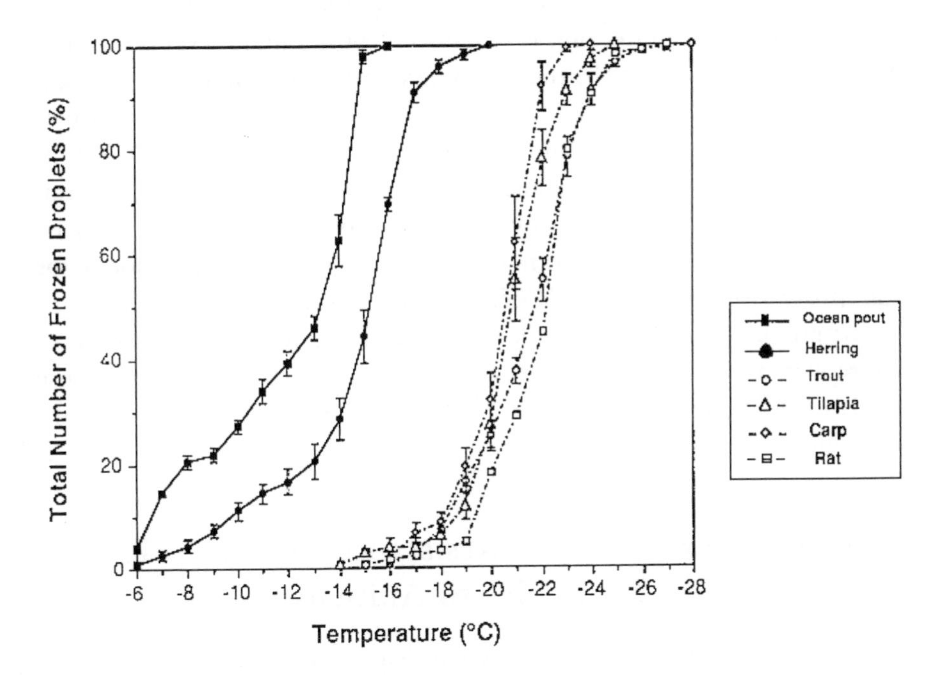

Fig. 4 Ice nucleation activity of serum samples from North Atlantic fishes, freshwater fishes and rat. For each replicate, freezing temperatures were determined in samples of 30 one-μl droplets by lowering the temperature on a cooling plate. Symbols are (■) ocean pout, (•) herring, (∘) rainbow trout, (Δ) tilapia, (◊) bighead carp, and (□) rat. Reprinted with permission from Verdier *et al.* (1996).[53]

suggest that the ocean pout has anti-ice Igs. In order to generate an immune response leading to the production of anti-ice Igs, ice crystals must have been present within the fish. This would refute the ice exclusion hypothesis presented above.

Taken together, these two contrasting hypotheses can be considered as two arms of a protective mechanism rather than alternative mechanisms of protection from freezing. A barrier at the skin would, when intact, prevent most ice from getting into the fish and, when a crystal would bypass the barrier, the high concentrations of antifreeze throughout body fluids would then stop crystal growth. It is clear that skin AF(G)Ps, whether intra- or extracellular, would never be sufficient to fully protect fish from freezing. However, their contribution to overall freeze protection is probably still very significant. Overall, this two-level protection from freezing is reminiscent of the multi-level protection against infection in fish. In the fish innate immune system, (1) the skin and the gut are barriers with antibodies, proteases, antimicrobial peptides, etc. that prevent most pathogens from getting into the fish, and (2) the humoral and cellular innate immunity within the fish can then neutralize or destroy any pathogen that bypasses the barriers. Using protection of fish from pathogens as a conceptual model for the protection of fish from destructive crystals may be helpful in determining the biological role of AF(G)P in fish freeze resistance. Moreover, as ice itself appears to generate an acquired immune response in fish leading to Ig production,[53] the analogy with pathogens seems biologically appropriate.

Perspectives on Freeze Resistance in Fish — From Molecules to Animals

AF(G)Ps appear to have evolved separately many times among teleost fishes.[4] This is the most logical explanation for the present diversity of these proteins. However, it is likely that there are many more AF(G)Ps and possibly even distinct new types as yet undiscovered among polar and cold temperate fish species. The search for AF(G)Ps has been guided mainly by assays for thermal hysteresis, which requires high AF(G)P concentrations. However, the activity of AF(G)Ps can be detected through changes in ice crystal morphology even when they are present at concentrations far too low to cause detectable hysteresis. Two fish species that were previously thought to have no antifreeze activity (GL Fletcher (1998), personal communication) have now shown clear evidence for AF(G)P presence when investigated using more sensitive methods.[54,55] Longhorn sculpin (*Myoxocephalus octodecimspinosis*) produces the type IV AFP but the protein is present at less than 10 μM in blood.[54] Therefore, activity was only detected after chromatography of blood proteins and concentration of the fractions.[54] Haddock (*Melanogrammus aeglefinus*) have trace levels of a Ca^{2+}-dependent AFP.[55] Electrophoretic analysis and activity measurements on haddock blood fractions from size-exclusion HPLC suggest an approximate M_r of 12–15,000 for this AFP (V Ewart (1999), unpublished results). The AFPs of haddock and longhorn sculpin are remarkable in that they are both from species closely related to those known to make distinct AF(G)P types, with AFGPs being the form in all other gadoids and type I AFPs the predominant form among sculpins.

The presence of trace-level AF(G)Ps in some fish species raises questions regarding the distributions and roles of AF(G)Ps in fish. These AF(G)Ps would not likely have been discovered if they were present in species with other abundant AF(G)Ps, such as were identified previously (Table 1). The haddock and longhorn sculpin AFPs are present at levels that would be easily masked by more abundant AF(G)Ps if both were present in a fish species. This raises the interesting possibility that multiple AF(G)Ps are present in some fish species but that only the most active have been identified because of the limitations of our thermal hysteresis-based assay method. The discovery of trace-level AF(G)Ps also supports the contention of Davies *et al.* (1993) that part of what gives a protein an antifreeze role in fish, assuming some level of ice-binding activity, is sheer abundance.[56] Taken further, this would suggest that the subset of proteins that can bind to ice in fish might be much larger than originally thought but that most could have been overlooked if they co-exist with other AF(G)Ps that are present at much higher concentrations.

Another interesting aspect of the trace-level AF(G)Ps is their role in fish. Why would fish produce an AF(G)P at levels that are far lower than required for protection from freezing? It is possible that some fish have lost their abundant AF(G)Ps due to a lack of selective pressure and have very low circulating levels as a result. This has been suggested as a possible reason for lower AF(G)P activity in yellowtail flounder (*Pleuronectes ferrugineus*) than in winter flounder[57] and it may sometimes be the case. However, in haddock, which have a trace-level AFP, there should be considerable selective pressure for an abundant AF(G)P as the species has succumbed to mass die-offs associated with freezing temperatures.[58] The alternative possibility is that AF(G)Ps play a different role in these fish species that is not freezing-related and that requires much lower protein concentrations for biological relevance.[54] AF(G)Ps have been shown to bind membrane phosopholipids *in vitro* and protect the lipid bilayer from cold disruption.[59] This may also be a natural role of these AF(G)Ps in some fish species or the AF(G)Ps might have altogether different roles that are, as yet, unknown.

Conclusions

Questions central to our understanding of the AF(G)Ps in fish, including how the proteins interact with ice, how they protect fish from freezing and what other roles, if any, they might play, must be addressed if we are to understand how fish are able to survive in freezing sea water. Moreover, many of the species that produce AF(G)Ps and depend on these proteins for their survival are harvested in the wild fisheries or are beginning to be farmed through aquaculture initiatives. Therefore, understanding of the AF(G)Ps has become important for both biological and economic reasons. The study of AF(G)Ps extends from the biophysics of molecular contacts at the ice surface to fish population ecology and this broad scope will be valuable in revealing the subtleties of these proteins and their roles in fish.

Acknowledgements

I thank Susan Douglas (NRC IMB) for helpful review of the manuscript.

References

1. Ewart KV, Lin Q and Hew CL (1999). Structure, function and evolution of antifreeze proteins. *Cell. Mol. Life Sci.* **55**: 271–283.
2. Goddard SV and Fletcher GL (1994). Antifreeze proteins: their role in cod survival and distribution from egg to adult. *ICES Mar. Sci. Symp.* **198**: 676–683.
3. Goddard SV, Morgan MJ and Fletcher GL (1998). Influence of plasma antifreeze glycoproteins on temperature selection by Atlantic cod (*Gadus morhua*) in a thermal gradient. *Can. J. Fish. Aquat. Sci.* **54**(Suppl.) 1: 88–93.
4. Scott GK, Fletcher GL and Davies PL (1986). Fish antifreeze proteins: recent gene evolution. *Can. J. Fish. Aquat. Sci.* **43**: 1028–1034.
5. Cheng CC and DeVries AL (1991). The role of antifreeze glycopeptides and peptides in the freezing avoidance of cold-water fish. In: di Prisco G (ed.) *Life Under Extreme Conditions.* Springer-Verlag, Berlin, pp. 1–14.
6. Chao H, Hodges RS, Kay CM, Gauthier SY and Davies PL (1996). A natural variant of type I antifreeze protein with four ice-binding repeats is a particularly potent antifreeze. *Protein Sci.* **5**: 1150–1156.
7. Scott GK, Davies PL, Shears MA and Fletcher GL (1987). Structural variation in the alanine-rich antifreeze proteins of the Pleuronectidae. *Eur. J. Biochem.* **168**: 629–633.
8. Hew CL, Joshi S, Wang N-C, Kao MH and Ananthanarayanan VS (1985) Structures of shorthorn sculpin antifreeze polypeptides. *Eur. J. Biochem.* **151**: 167–172.
9. Davies PL, Roach, AH and Hew CL (1982). DNA sequence coding for an antifreeze protein precursor from winter flounder. *Proc. Natl. Acad. Sci. USA* **79**: 335–339.
10. Chakrabartty A, Hew CL, Shears M and Fletcher GL (1988). Primary structures of the alanine rich antifreeze polypeptides from grubby sculpin (*Myoxocephalus aenaeus*). *Can. J. Zool.* **66**: 403–408.

11. Ewart KV, Rubinsky B and Fletcher GL (1992). Structural and functional similarity between fish antifreeze proteins and calcium-dependent lectins. *Biochem. Biophys. Res. Commun.* **185**: 335–340.
12. Ewart KV and Fletcher GL (1993). Herring antifreeze protein: primary structure and evidence for a C-type lectin evolutionary origin. *Mol. Mar. Biol. Biotechnol.* **2**: 20–27.
13. Ng NF, Trinh KY and Hew CL (1986). Structure of an antifreeze polypeptide from the sea raven (*Hemitripterus americanus*). *J. Biol. Chem.* **261**: 15690–15695.
14. Li XM, Trinh KY, Hew CL, Buettner B, Baenziger J and Davies PL (1985). Structure of an antifreeze polypeptide and its precursors from the ocean pout (*Macrozoarces americanus*). *J. Biol. Chem.* **260**: 12904–12909.
15. Scott GK, Hayes PH, Fletcher GL and Davies PL (1988). Wolffish antifreeze protein genes are primarily organized as tandem repeats that each contain two genes in inverted orientation. *Mol. Cell Biol.* **8**: 3670–3675; *J. Biochem.* **126**: 387–394 (1999).
16. Miura K, Ohgiya S, Hoshino T, Nemoto N, Odaira M, Nitta K and Tsuda S (1999). Determination of the solution structure of the N-domain plus linker of Antarctic eel pout antifreeze protein RD3. *J. Biochem.* **126**: 387–394.
17. Deng G, Andrews DW and Laursen RA (1997). Amino acid sequence of a new type of antifreeze protein, from the longhorn sculpin *Myoxocephalus octodecimspinosis. FEBS Lett.* **402**: 17–20.
18. Yeh Y and Feeney RE (1996). Antifreeze proteins: structures and mechanisms of function. *Chem. Rev.* **96**: 601–617.
19. Sönnichsen FD, Sykes BD and Davies PL (1995). Comparative modeling of the three-dimensional structure of type II antifreeze protein. *Protein Sci.* **4**: 460–471.
20. Ewart KV, Yang DSC, Ananthanarayanan VS, Fletcher GL and Hew CL (1996). Ca^{2+}-dependent antifreeze proteins: modulation of conformation and activity by divalent metal ions. *J. Biol. Chem.* **271**: 16627–16632.
21. Gronwald W, Loewen MC, Lix B, Daugulis AJ, Sönnichsen FD, Davies PL and Sykes BD (1998). The solution structure of type II antifreeze protein reveals a new member of the lectin family. *Biochemistry* **37**: 4712–4721.
22. Drickamer K and Taylor ME (1993). Biology of animal lectins. *Ann. Rev. Cell Biol.* **9**: 237–264.

23. Epstein J, Eichbaum Q, Sheriff S and Ezekowitz RAB (1996). The collectins in innate immunity. *Curr. Biol.* **8**: 29–35.
24. Natori S (1991). Dual functions of insect immunity proteins in defence and development. *Res. Immunol.* **141**: 938–939.
25. Ewart KV, Li Z, Yang DC, Fletcher GL and Hew CL (1998). The ice-binding site of Atlantic herring antifreeze protein corresponds to the carbohydrate-binding site of C-type lectins. *Biochemistry* **37**: 4080–4085.
26. Loewen MC, Gronwald W, Sönnichsen FD, Sykes BD and Davies PL (1998). The ice-binding site of sea raven antifreeze protein is distinct from the carbohydrate-binding site of the homologous C-type lectin. *Biochemistry* **37**: 17745–17753.
27. Kao MH, Fletcher GL, Wang NC and Hew CL (1986). The relationship between molecular weight and antifreeze polypeptide activity in marine fish. *Can. J. Zool.* **64**: 578–582.
28. Knight CA, DeVries AL and Oolman LD (1984). Fish antifreeze protein and the freezing and recrystallization of ice. *Nature* **308**: 295–296.
29. Hon W-C, Griffith M, Mlynarz A, Kwok YC and Yang DSC (1995). Antifreeze proteins in winter rye are similar to pathogenesis-related proteins. *Plant Physiol.* **109**: 879–889.
30. Knight CA, Cheng C-HC and DeVries AL (1991). Adsorption of α-helical antifreeze peptides on specific ice crystal surface planes. *Biophys. J.* **59**: 409–418.
31. Knight CA, Driggers E and DeVries AL (1993). Adsorption to ice of fish antifreeze glycopeptides 7 and 8. *Biophys. J.* **64**: 252–259.
32. DeLuca CI, Davies PL, Ye Q and Jia Z (1998). The effects of steric mutations on the structure of type III antifreeze protein and its interaction with ice. *J. Mol. Biol.* **275**: 515–525.
33. Chao H, DeLuca CI and Davies PL (1995). Mixing antifreeze protein types changes ice crystal morphology without affecting antifreeze activity. *FEBS Lett.* **357**: 183–186.
34. Grandum S, Yabe A, Nakagomi K, Tanaka M, Takemura F, Kobayashi Y and Frivik P-E (1999). Analysis of ice crystal growth for a crystal surface containing adsorbed antifreeze proteins. *J. Cryst. Growth* **205**: 382–390.
35. Sicheri F and Yang DS (1995). Ice-binding structure and mechanism of an antifreeze protein from winter flounder. *Nature* **375**: 427–431.
36. Petrenko VF (1994). *The Surface of Ice.* CRREL Report 94–22, p. 36.

37. Wen D and Laursen RA (1992). A model for binding of an antifreeze polypeptide to ice. *Biophys. J.* **63**: 1659–1662.
38. Sönnichsen FD, DeLuca CI, Davies PL and Sykes BD (1996). Refined solution structure of type III antifreeze protein: hydrophobic groups may be involved in the energetics of the protein-ice interaction. *Structure* **4**: 1325–1337.
39. Chao H, Houston ME Jr, Hodges RS, Kay CM, Sykes BD, Loewen MC, Davies PL and Sönnichsen FD (1997). A diminished role for hydrogen bonds in antifreeze protein binding to ice. *Biochemistry* **36**: 14652–14660.
40. Haymet ADJ, Ward LG, Harding MM and Knight CA (1998). Valine substituted winter flounder "antifreeze": preservation of ice growth hysteresis. *FEBS Lett.* **430**: 301–306.
41. Haymet ADJ, Ward LG and Harding MM (1999) Winter flounder "antifreeze" proteins: synthesis and ice growth inhibition of analogues that probe the relative importance of hydrophobic and hydrogen-bonding interactions. *J. Am. Chem. Soc.* **121**: 941–948.
42. Harding MM, Ward LG and Haymet ADJ (1999). Type I "antifreeze" proteins: structure-activity studies and mechanisms of ice growth inhibition. *Eur. J. Biochem.* **264**: 653–665.
43. Baardsnes J, Kondejewski LH, Hodges RS, Chao H, Kay C and Davies PL (1999). New ice-binding face for type I antifreeze protein. *FEBS Lett.* **463**: 87–91.
44. Alghren JA, Cheng C-H C, Schrag JD and DeVries AL (1988). Freezing avoidance and the distribution of antifreeze glycopeptides in body fluids and tissues of Antarctic fish. *J. Exp. Biol.* **137**: 549–563.
45. O'Grady SM, Ellory JC and DeVries AL. (1982). Protein and glycoprotein antifreezes in the intestinal fluid of polar fishes. *J. Expt. Biol.* **98**: 429–438.
46. Fletcher GL, King MJ, Kao MH and Shears MA (1989). Antifreeze proteins in the urine of marine fish. *Fish. Physiol. Biochem.* **6**: 121–127.
47. Valerio PF, Kao MH and Fletcher GL (1992). Fish skin: an effective barrier to ice crystal propagation. *J. Expt. Biol.* **164**: 135–151.
48. Schneppenheim R and Theede H (1982). Freezing-point depressing peptides and glycopeptides from Arctic-boreal and Antarctic fish. *Polar Biol.* **1**: 115–123.
49. Davies PL, Fletcher GL and Hew CL (1989). Fish antifreeze protein genes and their use in transgenic studies. In: MacLean N (ed.) *Oxford*

Surveys on Eucaryotic Genes, Vol. 6. Oxford University Press, New York, pp. 85–109.

50. Gong Z, Ewart KV, Hu Z, Fletcher GL and Hew CL (1996). Skin antifreeze protein genes of the winter flounder, *Pleuronectes americanus*, encode distinct and active polypeptides without the secretory signal and prosequences. *J. Biol. Chem.* **271**: 4106–4112.
51. Low WK, Miao M, Ewart KV, Yang DSC, Fletcher GL and Hew CL (1998). Skin-type antifreeze protein from the shorthorn sculpin, *Myoxocephalus scorpius*: expression and characterization of a M_r 9,700 recombinant protein. *J. Biol. Chem.* **273**: 23098–23103.
52. Kam M, Perl-Treves D, Caspi D and Addadi L (1992). Antibodies against crystals. *FASEB J.* **6**: 2608–2613.
53. Verdier J-M, Ewart KV, Griffith M and Hew CL (1996). An immune response to ice crystals in North Atlantic fishes. *Eur. J. Biochem.* **241**: 740–743.
54. Deng G and Laursen RA (1998). Isolation and characterization of an antifreeze protein from the longhorn sculpin, *Myoxocephalus octodecimspinosis. Biochim. Biophys. Acta* **1388**: 305–314.
55. Ewart KV, Blanchard B, Johnson SC, Bailey WL, Martin-Robichaud DJ and Buzeta MI (2000). Freeze susceptibility in haddock (*Melanogrammus aeglefinus*). *Aquaculture*, in press.
56. Davies PL, Ewart KV and Fletcher GL (1993). The diversity and distribution of fish antifreeze proteins: new insights into their origins. In: Hochachka PW and Mommsen TP (eds.) *Biochemistry and Molecular Biology of Fishes*, Vol 2. Elsevier Science Publishers, Amsterdam, pp. 279–291.
57. Scott GK, Davies PL, Shears MA and Fletcher GL (1987). Structural variation in the alanine-rich antifreeze proteins of the Pleuronectidae. *Eur. J. Biochem.* **168**: 629–633.
58. Templeman W (1965). Mass mortalities of marine fishes in the Newfoundland area presumably due to low temperature. *Spec. Publ. Int. Comm. N.W. Atlant. Fish.* **6**: 137–147.
59. Hays LM, Feeney RE, Crowe LM, Crowe JH and Oliver AE (1996). Antifreeze glycoproteins inhibit leakage from liposomes during thermotropic phase transitions. *Proc. Natl. Acad. Sci. USA* **93**: 6835–6840.

Chapter 4

Origins and Evolution of Fish Antifreeze Proteins

Christina C-M Cheng
Department of Animal Biology
University of Illinois at Urbana-Champaign
515 Morrill Hall, 505 South Goodwin Ave.
Urbana, IL 61801, USA

Arthur L DeVries
Department of Molecular and Integrative Physiology
University of Illinois at Urbana-Champaign
524 Burrill Hall, 407 South Goodwin Avenue
Urbana, IL 61801, USA

Introduction

The Antarctic and Arctic Oceans, and the near-shore waters of the north Atlantic and Pacific Oceans during winter months, delimit the cold extremes for marine life. These waters share a common physical property: they are at or near their freezing point and frequently ice-laden, which in combination represent a formidable challenge to the survival of ectothermic marine teleost fishes. The basis for this is simple — the colligative solute contents of the body fluids of marine bony fishes[1,2] including cold-adapted polar species[3,4] are substantially lower than seawater, about 300 to 600 mOsM versus 1000 mOsM, which means the fishes have a higher freezing point, about −0.56°C to −1.1°C versus −1.9°C of seawater. The fishes are therefore supercooled with respect to freezing seawater, and by these physical terms alone, they could not avoid freezing indefinitely, especially when ice is present, because upon contact, ice growth will rapidly propagate throughout the supercooled body fluids leading to organismal freezing. Unlike some other vertebrate ectotherms, such as hibernating frogs and turtles which tolerate extensive freezing and recover fully upon thawing,[5,6] even partial freezing is lethal for marine teleosts.[7]

Some teleosts utilize behavioral strategies to avoid freezing, by migrating off-shore to warmer water such as the longhorn sculpin,[8] or occupying deep, ice-free water as in the case of some species in the fjords of Labrador.[7] Another strategy is to lower the blood colligative freezing point by elevating blood electrolyte and osmolyte concentrations. The rainbow smelt *Osmerus mordax*, produces as much as 400 mOsm serum glycerol at −2°C, rendering them near-isosmotic with seawater.[9]

Other polar and subpolar teleost fishes are restricted to freezing icy habitats because of environmental constraints or life history. For instance, the Antarctic fish fauna is largely confined within the frigid water of the Southern Ocean due to the massive Antarctic Circumpolar Current which is both an oceanographic barrier against migration, and a thermal barrier that precipitated and sustained the freezing of the Antarctic water.[10,11] A number of fish taxa that occupy frigid habitats overcame the threat of freezing through an innovative biological solution — they evolved novel ice-binding antifreeze proteins[12,13] which enabled them to successfully colonize icy habitats that were otherwise out of their reach. Antifreeze proteins are primarily synthesized in the liver and secreted into the circulation. Unlike colligative solutes which act to lower the freezing point of water so ice does not form, antifreeze proteins act by recognizing and binding to ice crystals that enter the fish, and arrest ice growth within the temperatures the fish live in, thus preserving their body fluids in the liquid state. The mechanism is non-colligative, called adsorption-inhibition,[14] and is a subject widely investigated in the field of antifreeze research (see other chapters and the references therein).

Cool Ocean, Hot Bed of Antifreeze Evolution

Freezing death means termination of an organismal lineage, which would impose great incentive or selective pressure on an organism to evolve a protective function. The evolution of fish antifreeze proteins (AF(G)Ps) eminently exemplifies evolutionary creativity and propensity

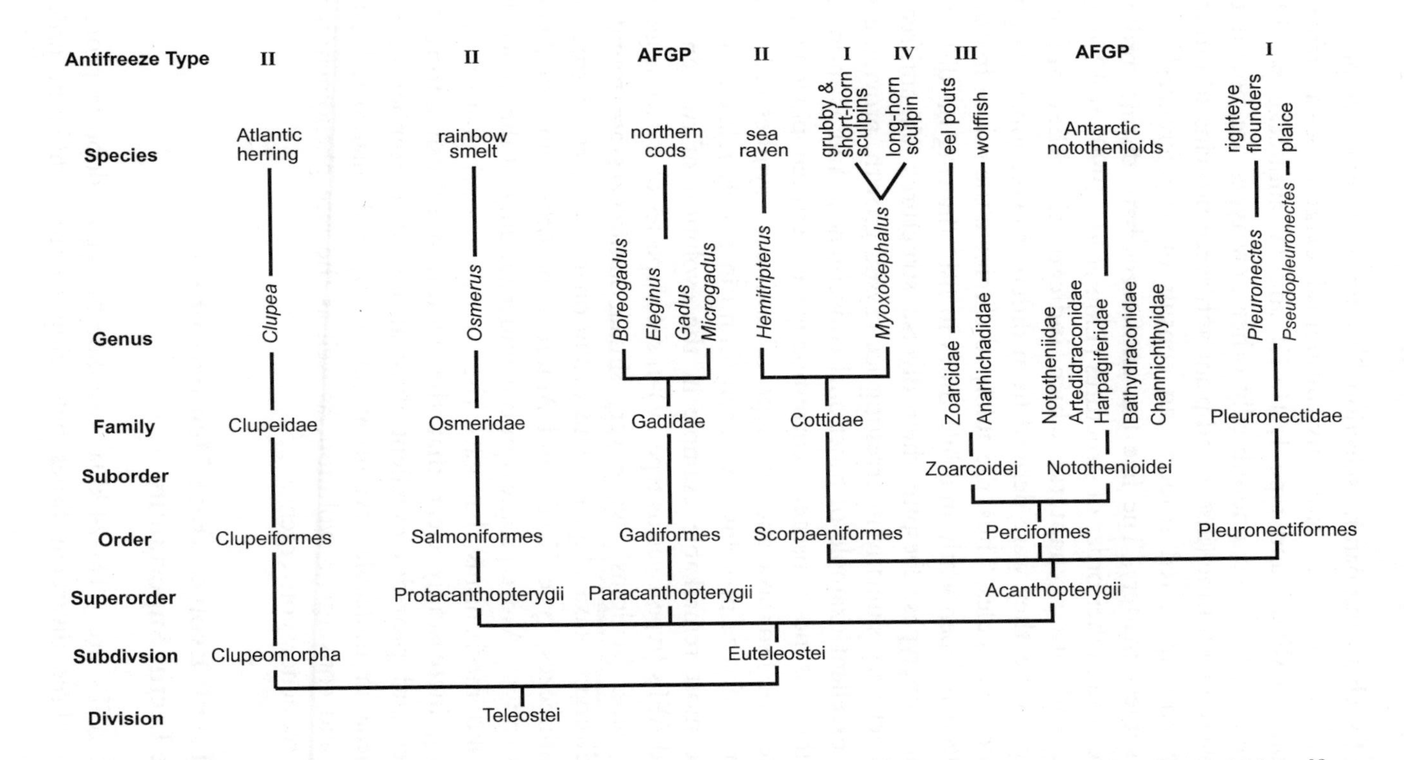

Fig. 1 Types of AF(G)Ps in marine teleost fish and the phylogenetic relationship among the taxa according to Nelson.[19]

in face of such strong natural selection. Five types of antifreeze proteins have evolved to meet this environmental exigency — antifreeze glycoprotein (AFGP) and type I, II, III and IV antifreeze peptides (AFP) (Fig. 1),[12,13] and undoubtedly other new types are bound to be discovered with time (insect and plant antifreeze proteins add to this diversity, and interested readers can consult other references[12,15–18] on the subject matter). The five fish antifreezes have distinct protein sequences and higher order of structures (Liou *et al.* (1999) and Nelson (1994), and other chapters) which indicate that they are not homologous, i.e. they were derived from different evolutionary origins or ancestral molecules. Thus, the ability to eliminate the risk of freezing has evolved independently multiple times in fish, drawing on different ancestral archetypes. Despite their diverse structures, all antifreeze proteins act very similarly in arresting ice crystal growth, providing us with an excellent example of functional convergence. And as shown in Fig. 1, there is no consistent correspondence between phylogenetic relatedness of fish taxa and the type of antifreeze synthesized. Very different antifreeze proteins have been found in closely related species, with the most remarkable example in the evolution of two entirely unrelated AFPs (type I versus type IV) in sister species of the same genus *Myoxocephalus* (sculpins), while very similar antifreezes are found in highly divergent taxa (Fig. 1). The evolutionary ancestry of three of these antifreezes, type II, IV and Antarctic notothenioid fish AFGP are now known. As we progress in our understanding of the molecular origins and mechanisms that gave rise to these novel proteins, it is becoming increasingly clear that the evolution of fish antifreezes embodies and provides excellent illustrations of a number of key phenomena in molecular evolution, including molecular parallelism, *de novo* and convergent evolution, as well as the rare phenomenon of protein sequence convergence.

Type II AFP Evolved from Members of the C-Type Lectin Superfamily

Type II AFPs are ~15–24 kDa, disulfide-bridged, globular proteins found in three divergent fishes belonging to three different orders in two different subdivisions — the rainbow smelt *Osmerus mordax* (order Osmeriformes), herring *Clupea harengus harengus* (order Clupeiformes)[20,21] and a cottid, the sea raven *Hemitripteris americanus* (order Scorpaeniformes)[22] (Fig. 1). Herring and smelt AFPs are closely related, sharing about 85% amino acid identity. Smelt AFP contains an N-linked carbohydrate which does not appear to be involved in the antifreeze function.[23] Sea raven AFP is more divergent, with about 40% identity to herring and smelt AFP. All three AFPs show about 30% amino acid identity to the carbohydrate recognition domain (CRD) of C-type (Ca^{2+}-dependent) lectins, suggesting these AFPs descended from the latter.[21,23]

C-type lectins comprise a large superfamily of proteins, some of which bind specific sugars or sugar moieties when complexed with Ca^{2+} in the binding site. Others contain C-type CRD-like sequence motif and exhibit similar protein fold but serve functions other than carbohydrate binding, and are therefore more appropriately named C-type lectin like domains or CTLDs.[24] The type II AFPs in fact can be considered as CTLDs based on sequence identity and similar global fold[25,26] but with an ice-binding rather than sugar-binding function. The herring and smelt AFPs require Ca2+ for antifreeze activity,[21,23] reminiscent of the Ca^{2+} requirement of its putative CRD progenitor, and supporting their evolutionary homology. Mutagenesis studies indicated that the ice-binding site of herring AFP may in fact correspond to the carbohydrate-binding site in the C-type lectin mannose-binding protein.[25]

The NMR solution structure of sea raven AFP shows a global fold very similar to mannose-binding protein and the CTLD lithostatin (pancreatic stone protein),[26] again supporting a lectin origin. Sea raven AFP is independent of Ca^{2+} for its activity,[23] and its NMR structure suggests that the loop region where the Ca^{2+} binding site resides in CRD homologs may be sufficiently different to preclude Ca^{2+} binding.[24,26] In contrast to herring and smelt AFP, mutagenesis studies indicated that the ice-binding surface of sea raven AFP resides outside of the counterpart of the Ca^{2+} and sugar binding motif.

The sequence and structural similarities between type II AFPs and CRD of C-type lectins or CTLDs established quite conclusively that the former arose from a lectin in the fish, presumably through duplication of a pre-existing lectin gene, and subsequent sequence divergence lead to the acquisition of the new ice-binding function. It is not quite as clear whether the type II AFPs from all three taxa evolved from the same C-type lectin progenitor and the sea raven AFP diverged to the extent of losing the Ca^{2+} dependence, or the 85% identical herring and smelt AFPs share a common ancestor, and the more divergent sea raven AFP was derived independently from a separate CTLD. The evidence that the putative ice-binding surface of sea raven AFP is distinct from that of the herring and smelt tend to argue for a separate lectin antecedent.

An Apolipoprotein Ancestry is Inferred for Type IV AFP

Type IV AFP from the longhorn sculpin is the most recent addition to the fish AFPs.[27] It is a 108-residue (~12 kDa) protein and shares about 20% sequence identity with members of the exchangeable apolipoprotein superfamily. Apolipoproteins are the protein components of plasma lipoproteins which are macromolecular complexes responsible for lipid binding and transport and other related functions.[28] The three-dimensional structure of type IV is not yet available to demonstrate whether there is structural homology to the apolipoprotein to support the evolutionary relatedness between the two proteins despite low sequence similarity. However, there are distinct features in the primary sequence that lends support to an apolipoprotein ancestry.

Vertebrate apolipoproteins have long been recognized to contain multiple 22-residue repeats (encoded by the last exon — exon 4) each beginning with a single conserved proline residue, and each 22-mer repeat is comprised of two characteristic 11-mers in tandem.[28–30] Aligning the longhorn sculpin AFP with human apolipoprotein E (ApoE) and A-I (ApoA-I), residues 28 to 104 of the AFP were found to conform to the 22-mer repeat structure, and three of the possible four repeats begin with the characteristic single Pro.[12] This degree of Pro conservation in the AFP is closer to ApoA-I than ApoE. The three-dimensional crystal structures of human ApoA-I[31] and ApoE[32] have been solved. In human ApoA-I, the repeated 22-residue region comprises the lipid binding domain whose structure is a pseudo-continuous amphipathic helix punctuated by kinks at the regularly spaced Pro,[31] while in ApoE it comprises the LDL-receptor binding domain and has an antiparallel four-helix bundle structure.[32] It will be very interesting to see whether type IV AFP is structurally more similar to ApoA-I lipid binding domain than to the ApoE LDL-receptor binding domain. If this proves to be so, together with the primary structural similarity, it would be strong evidence that the longhorn sculpin type IV AFP evolved from the lipid-binding domain of the teleost's homolog of human ApoA-I.

The molecular mechanism of the evolution of type II and IV AFPs is common to how most new protein genes arose — by duplication of the gene sequence of a pre-existing protein or protein domain, followed by sequence divergence leading to the acquisition of a new function.[33,34] Their evolutionary ancestries can therefore be inferred through statistically meaningful sequence similarities to existing protein gene sequences in the database, which can be further supported by three-dimensional structural homology. Not all antifreezes however, have sequence similarity to currently known genes in the database. Such is the case of type I and type III AFPs, and northern cod AFGPs (see below), and thus their ancestries remain unsolved.

Evolution of AFGPs in Antarctic Notothenioids Caught in the Act

Unlike the evolution of type II and IV AFPs, which likely involved straightforward gene duplication and sequence divergence, the evolution of antifreeze glycoprotein (AFGP) proved to be more complex. AFGP was the first protein antifreeze discovered, three decades ago by DeVries[35] in the Antarctic notothenioid fish. It has the peculiar feature of being composed entirely of repeats of a very short sequence, a simple Thr-Ala-Ala monomer with each Thr residue O-linked to a disaccharide (left box). The repetitive protein sequence suggests that the encoding gene arose from *de novo* duplications of a short coding element for the tripeptide monomer. Although simple in primary structure, the

AFGPs are complex in protein heterogeneity (size and composition). Eight different sizes of AFGPs were initially identified and named AFGP 1–8. AFGP1 is the largest (34 kDa, n ≈ 55), and AFGP8 is the smallest (2.6 kDa, n = 4).[36,37] This nomenclature has persisted although better protein resolution techniques have now revealed the presence of many other intermediate sizes in different notothenioid species.[38] These length isoforms also contain minor compositional variations, in the replacement of some of the first Ala in the tripeptide repeats by Pro.[38–40] And unlike most proteins, which are encoded as a single molecule per gene, AFGP isoforms are not encoded by individual genes, but as distinct copies within large polyprotein genes. The large AFGP polyprotein precursor encoded in each gene comprises many AFGP molecules (as many as 41) linked in tandem by conserved three-residue spacers (mostly Leu-Asn-Phe or Leu-Ile-Phe), and post-translational removal of the spacers produce the individual mature AFGPs.[40–43]

The evolutionary origin of the structurally peculiar notothenioid AFGP protein and gene remained obscure until recently, after the technical difficulties of sequencing the highly repetitive AFGP polyprotein coding region were resolved and complete AFGP gene structures were obtained.[41,42] The AFGP gene was found to be evolutionary related to a trypsinogen-like protease from Atlantic plaice *Pleuronectes platessa* in the database[44] through sequence similarity (>70%) between the AFGP gene's 3′ flanking sequences and the coding sequence of the either the N- or C-terminus of the protease cDNA. However, it was not obvious how an AFGP gene could be derived from a trypsinogen gene since the amino acid sequences of the two proteins bear no resemblance whatsoever, indicating that molecular processes other than simply gene duplication were involved.

By cloning both the trypsinogen-like protease (TLP) and AFGP genes and cDNAs from the same notothenioid fish *Dissostichus mawsoni*, and comparative analyses of their sequences, a plausible molecular mechanism by which the TLP to AFGP transformation took place was deduced.[42] The protease gene has six exons and five introns (Fig. 2b), while AFGP gene has two exons and a single intron (Fig. 2a). There are three regions of high nucleotide sequence similarities between the two

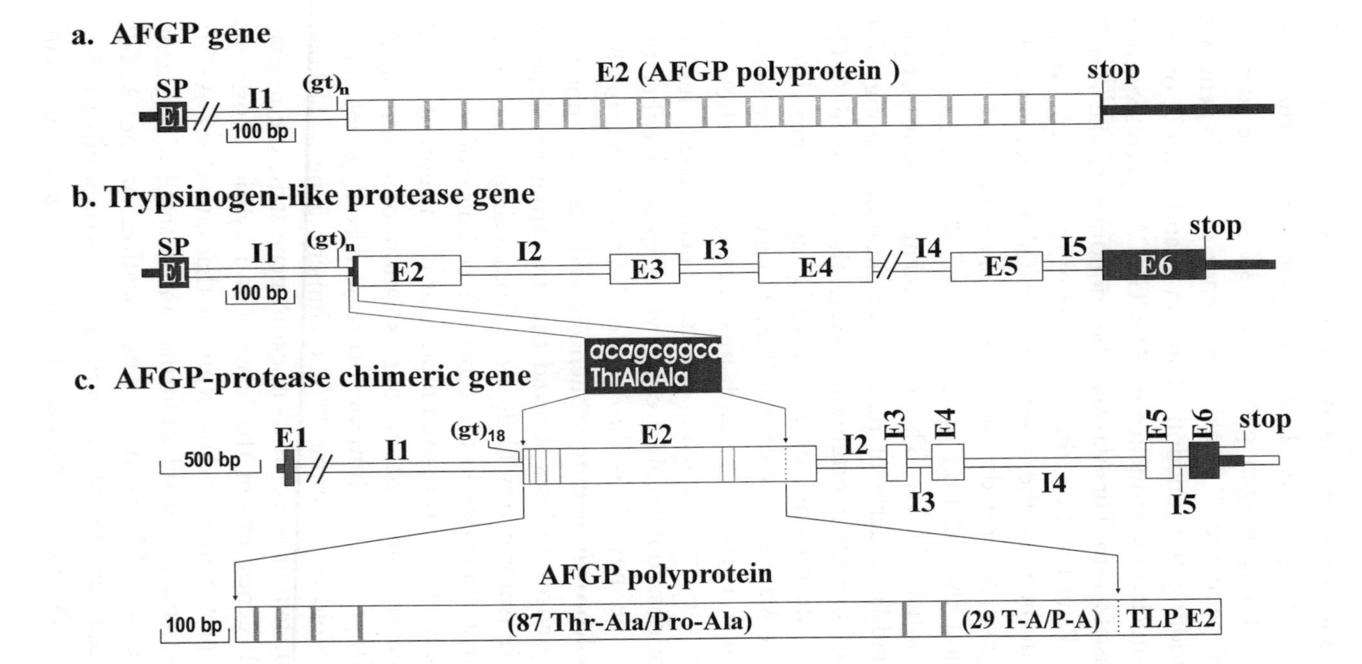

Fig. 2 Structure of a typical AFGP polyprotein gene **(a)** and trypsinogen-like protease (TLP) gene **(b)** from notothenioid fish. **(c)** Key evolutionary intermediate — a chimeric AFGP-protease gene from the notothenioid *D. mawsoni* in which a large AFGP polyprotein coding region was established but has not detached from the protease gene. Notothenioid AFGP gene arose from recruitment of the front (E1 and I1) and the tail (E6) segments of an ancestral protease gene, ***de novo*** creation of the AFGP coding region by iterative duplications of a nine-nucleotide Thr-Ala-Ala-coding element spanning the protease gene's I1-E2 junction (enlarged between b and c), and deletion of the bulk of the protease sequence (rest of E2 through I5). Reprinted with permission from Cheng and Chen (1999).

genes. Exon 1 of both gnes encode the signal peptide, and are 96% identical. The large (~2 kbp) intron 1 of AFGP gene contains the entire ~0.2 kbp intron 1 of trypsinogen gene with 93% sequence identity. Exon 6 of the TLP gene and 3′ flanking sequence of AFGP gene are 94% identical. Straddling the junction of intron 1 and exon 2 of the TLP gene is a nine-nucleotide sequence ACAGCGGCA (splice sequence in italics) that translate into Thr-Ala-Ala, the monomer of the AFGP tripeptide (Fig. 2b). Plausibly then, the formation of the incipient AFGP gene involved recruitment of the front and the tail segments of the ancestral protease gene, to provide for the secretory signal and downstream sequence, and the deletion of the bulk of the protease gene (remainder of E2 through I5). The Thr-Ala-Ala coding element that was positioned in the middle underwent repeated duplications (along with a spacer sequence of unknown origin) to create the repetitive AFGP polyprotein proper. The presence of a long $(gt)_n$ repeats immediately ahead of the Thr-Ala-Ala coding element (Fig. 2) very likely facilitated the first duplication through strand mispairing or replication slippage.[45,46] When the environmental selection heightened, further Thr-Ala-Ala duplications could rapidly occur via slippage at either the $(gt)_n$ sequence or the new Thr-Ala-Ala duplicants,[46] and further elongation of the polyprotein coding sequence could occur by unequal crossing over.[47] Sequences analyses showed that the proline residue in Pro-containing AFGP isoform very likely arose from a single-nucleotide substitution of the first Ala codon in the tripeptide coding sequence.[42,46]

The protease origin and the proposed molecular mechanism of AFGP evolution were recently validated by our isolation of evolutionary intermediates in the form of chimeric protease-AFGP genes from genome of notothenioid fish genome.[40] The chimeric gene (~11 kbp) isolated from *D. mawsoni* (Fig. 2c) is identical in gene structure to that of independent TLP genes, except that exon 2 is a hybrid, consisting of a large 5′ AFGP polyprotein coding segment that encodes 7 AFGP molecules (two large and five small isoforms) and a small 3′ segment similar in sequence to TLP exon 2. The location of the AFGP polyprotein coding segment in the chimeric gene corresponds to that of the single Thr-Ala-Ala-coding element in the TLP gene (element straddles I1-E2 junction), which strongly indicates that the repetitive AFGP polyprotein coding sequence arose from expansion of the element through iterative duplications. The chimeric protease-AFGP gene is not a pseudogene but transcriptionally active, as we have isolated and sequenced a partial chimeric protease-AFGP cDNA.[40]

In sum, the notothenioid AFGP gene evolved through a combination of recruitment of segments of an existing TLP gene, plus *de novo* amplification of a short Thr-Ala-Ala coding element to create an entirely new protein coding region for the new ice-binding function. The identification of a key evolutionary intermediate of this process in the form of chimeric genes in which the ancestral protease and fully expanded AFGP polyprotein coding regions are still attached provided validation of the molecular mechanism of the protease-to-AFGP transformation, as well as a rare view of the evolution of a novel protein gene in action.

Evolution of Arctic Cod AFGP by Protein Sequence Convergence

Unrelated to the notothenioids, a number of northern cod species in the family Gadidae also synthesize a family of AFGP isoforms that are nearly identical in primary structure to the Antarctic version and similar in protein heterogeneity. The major difference is that the Thr residues in the tripeptide repeats are occasionally replaced by an Arg.[48,49] Given that it is highly improbable for two very similar sets of proteins to have evolved independently by chance, the logical assumption would be that the two fish taxa inherited their AFGP gene from a common ancestor. However, this is at odds with their phylogenetic distance and their geographic origins. Notothenioids are modern Perciformes teleosts in the superorder Acanthopterygii, while cods are more basal, belonging to the order Gadiformes in the superorder Paracanthopterygii (Fig. 1).[19] Gadiform fishes are believed to have evolved in the boreal Atlantic where their fossils abound,[50] and the known AFGP-bearing gadids are exclusively north subpolar and polar in distribution.[51,52] Notothenioids are believed to have evolved *in situ* in the Antarctic waters and the modern notothenioid fauna is predominantly confined to the Southern Ocean, with no members at all north of the sub-Antarctic regions.[3,53] In addition, the north and south polar regions also had chronologically different glacial histories,[11,54] indicating that the need for antifreeze evolution would occur at different times for the two fish taxa.

We recently obtained the first gadid AFGP gene sequence, from the polar cod, *Boreogadus saida*.[43] Comparative sequence analyses with notothenioid AFGP genes strongly support a separate molecular origin for the cod AFGP gene. *B. saida* AFGP genes also have a polyprotein structure like the notothenioid AFGP genes in which multiple copies of AFGP coding sequences are linked by small cleavable spacers, but that is where the similarity ends. Several molecular evidence strongly indicate that the cod and notothenioid AFGP genes are not homologous: (i) Signal peptide sequences are not homologous. Notothenioid AFGP gene inherited and utilizes a TLP's signal peptide, while Arctic cod's signal peptide is distinct and does not match any current sequences in the database. (ii) Spacer sequences that linked the encoded AFGP molecules in the polyprotein are different, invoking different mechanisms of processing of the polyprotein precursors. The notothenioid spacers are the highly conserved tripeptide Leu/Phe-Ile/Asn-Phe, cleavable by a chymotrypsin-like protease, while in the Arctic cod they are either a single Arg or Arg-Ala-Ala-Arg, cleavable by a trypsin-like protease. The Arg is invariably coded by AGA, and always replaces a Thr in the tripeptide, and thus very likely arose from a single-base transversion (C to G) converting a Thr (ACA) in the AFGP tripeptide repeat to Arg (AGA) which became recruited as a spacer site. Apparently not all the Arg residues in the polyprotein are removed during post-translational processing, leading to the occasional Arg for Thr substitution observed in the mature AFGPs.[48,49] (iii) Codon bias of the nine-nucleotide sequence for the AFGP tripeptide repeats are distinctly different. In the notothenioids, the extant Thr-Ala-Ala codon bias reflects the sequence of its putative nine-nucleotide ancestral sequence in the protease progenitor (Fig. 2). It follows then that the distinct tripeptide codon bias of cod likely resulted from amplification of a different ancestral element with a different permutation of codons for the same three amino acids. (iv) Genomic loci of AFGP gene origin are different.

Although the genomic origin of Arctic cod AFGP gene is unknown at this time, clearly it did not evolve from a trypsinogen-related protease, because it shares no similarity with the latter, nor to any known sequence in the database thus far.[43] Thus, the cod AFGP gene was derived from a different evolutionary origin, and the two unrelated fish taxa arrived at very similar AFGPs through convergent evolution.[43]

The cod and notothenioid AFGPs in fact represent a definitive and rare example of convergent evolution at the protein sequence level, where two entire proteins are near-identical. One other instance where protein sequence convergence very likely occurred is the fibrillar proteins of spider dragline silk,[55] silkmoth chorion,[56] cockroach oothecin,[57] and lamprin — the annular cartilage protein in lamprey.[58] Singlets or tandem repeats of the pentapeptide Gly-Gly-Leu-Gly-Tyr deemed important in conferring the fibrillar structural property comprise segments of these proteins which as a whole are presumed non-homologous.[58] The AFGPs and the fibrillar proteins demonstrate that protein sequence convergence, though rare,[59] can occur. In both cases, it is likely made possible by the short sequence of the monomer (nine or 15 nucleotides), which has a significant statistical probability of occurring at multiple genomic locations, and the ease for short repeats to undergo expansion once the first duplication occurred. The selection of a permutation of codons from different genomic sites that will provide the same protein sequence presumably is shaped by the structural requirement necessary for the protein function in each respective case.

Causal Link Between Paleogeography, Evolution of Antifreeze Genotype and Phenotype, and Organismal Success

How genotypes produce phenotypes is often obscure because of the complexity of biological systems where multigenic traits is the rule. There are only a few clear examples where the evolution of or change in a genotype and the manifestation of the phenotype shows direct one-to-one correspondence. These include lens crystallin genes where over-expression of the crystallin proteins in the eye forms the light diffractive lens,[60] and the vertebrate opsin genes which encode chromophore-coordinating visual proteins that absorb optimally at specific wavelengths (red, blue or green) leading to color vision.[61,62] The AF(G)Ps constitute a prominent example among this select few. The evolution of the AF(G)P genes and encoded proteins are directly responsible for the physical retardation of ice growth in the fish in icy subzero temperatures, thus conferring the freeze avoidance phenotype and survival of the marine teleost fish. The adaptive value of the evolution of this protective function is amply clear — survival fitness and continuation of an organismic lineage. Additionally, the environmental selection that drove the evolution of antifreeze function in fish is also clear — the glaciation of the polar and subpolar oceans over geologic times. There are perhaps no other examples of adaptive evolution by natural selection that illustrate the connection between micro- and macro-evolution more clearly than the fish antifreezes.

The causal links between paleogeography, antifreeze evolution and organismal success are most clearly borne out in the AFGP-bearing Antarctic notothenioids. The Notothenioidei are largely endemic to the Antarctic region of the Southern Ocean, and it reigns as the predominant taxon of the Southern Ocean ichthyofauna.[3] With over 120 species currently described, the notothenioids represent 45% of the ~270 species of fish in the Southern Ocean.[63] The numerical dominance pales against their biomass; over 90% of fish catches on the shelves and upper slopes are notothenioids.[3] The members of the five endemic Antarctic families — Nototheniidae, Artedidraconidae, Harpagefieridae, Bathydraconidae and Channichthyidae (Fig. 1) — exhibit a diverse array of morphology, physiology and lifestyles, and occupy all niches throughout the water column. This adaptive organismal radiation and diversification is linked to two major factors — the isolation of the Antarctic environment, and the evolution of their AFGPs.

The paleogeograhic process that lead to the isolation of Antarctica began with the fragmentation of Gondwanaland during late Triassic to Early Jurassic about 160 million years ago (mya), which in turn commenced the development of paleoseaways around Antarctica (Fig. 3).[11,64] By about 36 mya, the last southern continent, South America, detached from Antarctica, opening up the Drake Passage. As

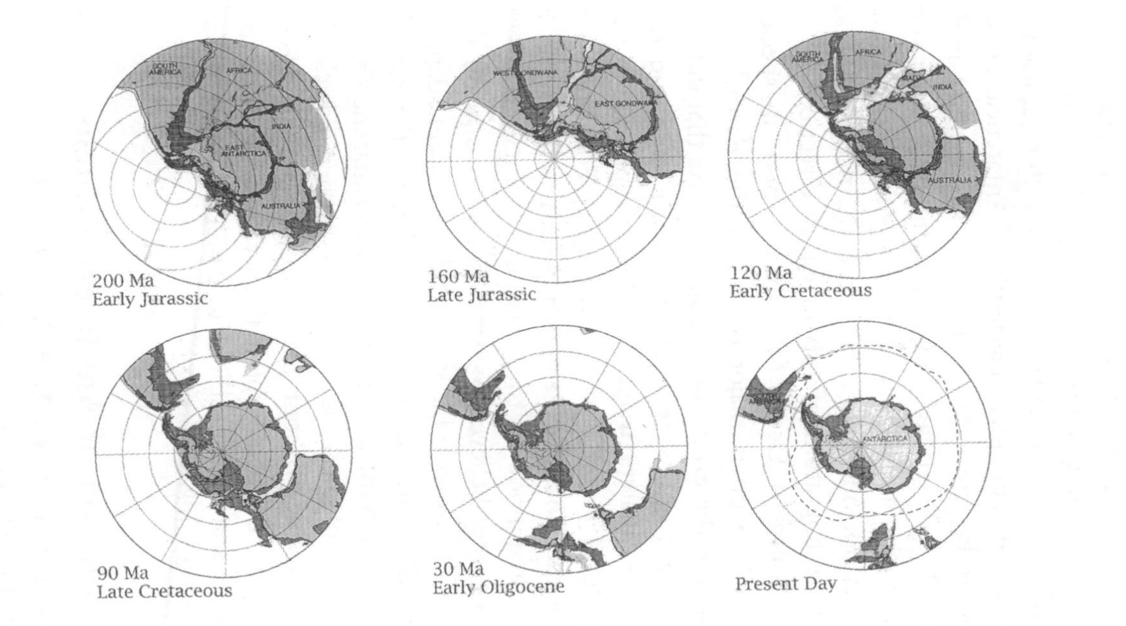

Fig. 3 Paleogeographic reconstruction of Gondwana from early Jurassic to early Oligocene, and present day map of Antarctica and the surrounding Southern Ocean. Fragmentation of southern Gondwana began at 160 Ma. Southern continents and Indian subcontinent detached and moved northward while Antarctica moved towards south polar position. South America was the last continent to detach, at about 36 Ma, opening up the Drake Passage. High latitude circumpolar flow began at 30 Ma. By 20 Ma, the vigorous Antarctic Circumpolar Current (ACC) was established, leading to thermal isolation and glaciation of Antarctica. The Antarctic Convergence (inclusive of ACC) in present day Southern Ocean is indicated by the dashed line. Antarctic notothenioids evolved within the Antarctic water, and today are largely confined south of the Convergence. Redrawn from Lawver *et al.* (1992).[64]

Antarctica continued to move towards the south polar position, sea floor spreading commenced the circumpolar current flow around Antarctica. The unrestricted flow of the Antarctic Circumpolar Current (ACC) began around 25 mya, decoupling the warm subtropical gyres from the Antarctic water; the ACC was fully established by about 20 mya, leading to the thermal isolation and subsequent glaciation of Antarctica. Frigid conditions as we find today are believed to culminate around mid-Miocene (10–15 mya).[10,11,64] Today's ACC is a massive clockwise moving oceanic current spanning 200 to 1000 km wide and reaching the ocean floor,[65] which not only acts as a thermal barrier but also as a physical barrier against migration of fish in either direction.

The contemporary Southern Ocean fish fauna is poor in species diversity when compared to other major oceans of the Earth.[3] The cause has been attributed to the extinction of many fish taxa that did not evolve antifreeze protection upon the isolation of Antarctica and the subsequent freezing of the Antarctic water.[3,53,63] This is supported by the late Eocene (40 mya) fossil records at the La Meseta Formation on the Seymour Island of the Antarctic Peninsula, which indicate that a cosmopolitan fauna existed prior to the isolation of Antarctica.[66–68] The notothenioid fishes were believed to have evolved *in situ* on the Antarctic margin from a shallow benthic stock, and underwent adaptive radiation and diversification into the open niches left by the extinct groups, to become the predominant fish taxon in the Antarctic Ocean today.[3,53,63] However, the ecological success of the notothenioids clearly would not have been possible without the evolution of their AFGP in the first place, to enable them to exploit the icy freezing Antarctic environments. For this reason the notothenioid AFGP is considered as a key evolutionary innovation.[63]

Time of Evolution of Antifreeze Protein as Chronometer of the Advent of Polar Water Glaciation

The causal link between the chilling of the polar oceans as the environmental driving force and the evolution of the antifreeze protection means that the two events would also be temporally linked. Thus, if the time of emergence of the antifreeze gene can be determined, it can serve as a biological indicator of the earliest geologic time when the marine body of water inhabited by a particular antifreeze-bearing taxon approached or reached freezing. Estimating the time of emergence of a new gene requires knowing the progenitor and its sequence, so that the sequence divergence between the two genes can be determined, which in turn can be used to determine the time of their split using an appropriate molecular clock rate. In the case of the notothenioid AFGP where the progenitor is clearly identified, and the extant version of the progenitor gene sequence is obtained from the same fish, the amount of sequence divergence was determined.[42] Using a teleost (salmon) mitochondrial sequence divergence rate, we have estimated the time of split between the ancestral trypsinogen protease gene and the offspring AFGP gene to be about five to 14 mya.[42] This time estimate is consistent with the mid-Miocene (ten to 15 mya) onset of full-fledged Antarctic glaciation estimated by paleo-oceanographic methods.[10,11] Additionally, molecular phylogenetic analyses of notothenioid families using rRNA genes also point to a burst of phyletic diversification at about five to 15 mya.[69] The separate time estimates of the freezing of the Antarctic water, emergence of AFGP gene, and organismal diversification appear to converge satisfactorily at the same mid-Miocene time frame, lending confidence to these estimates. However, there are inherent caveats in the assumptions used in both the physical and molecular methods. In the case of the molecular method, the sequence divergence rate used in estimating the notothenioid protease-AFGP divergence time is from the unrelated salmon,[70] and it is based on mitochondrial DNA which in general evolves much faster than nuclear genes.[71] To better estimate the protease-AFGP split, a more realistic and robust calibration of the rate of sequence divergence of notothenioid nuclear genes is in order.

Nevertheless, using the divergence time of an antifreeze gene from its *bona fide* ancestral gene to infer the time of advent of glaciation of a particular body of marine water is theoretically possible and may even be useful in addressing the question of why such diverse types of antifreeze proteins (all the five known types actually) have evolved independently in the northern fishes. One possible scenario is that the cyclical glacial advances and retreats throughout the Quaternary Epoch since the initial northern glaciation in late Pliocene (~2.5 mya)[54] might have created freezing conditions in smaller scale marine environments in the northern oceans at different times. In other words, the freezing selection pressure might have be temporally and spatially disjoint such that, depending on where the teleost's habitat range was, closely related species might not have been subjected to selection simultaneously, or distantly related species might have been, thus leading to multiple independent evolution of the northern fish antifreeze. This hypothesis can be tested if the ancestral molecule of each antifreeze protein can be veritably identified, and the time of split between the two can be determined with sufficient resolution with respect to the times between glacial episodes.

Concluding Remarks

The glaciation of the polar and subpolar seas had driven independent evolution of diverse antifreeze proteins in teleost taxa to survive in icy seawater. The diversity of these proteins, and thus the diversity of their genetic origins provide excellent avenues for the study of molecular evolution of novel protein genes. Three of the five known types of fish antifreezes now have known ancestry. This numerical count does not completely reflect the multiple molecular origins from which they were derived. Antifreeze glycoprotein had evolved independently twice — in the Antarctic notothenioid fish from a trypsinogen-like serine protease, and in the unrelated northern cod from a yet to be determined origin which is definitively not a trypsinogen-like serine protease. Type II AFP has evolved independently three times in three unrelated taxa — from perhaps a homologous calcium-dependent lectin ancestor in the case of the AFP of smelt and herring, and from another C-type lectin-like domain (CTLD) more akin to pancreatic stone protein which is Ca^{2+}-independent in the case of the sea raven AFP. Notothenioid AFGP coding sequence arose by *de novo* duplications of a rudimentary Thr-Ala-Ala coding element, and the northern cod version very likely arose by an comparable mechanism but from a different ancestral short sequence, to arrive at the same protein by convergent evolution. Type II AFP, as well as type IV evolved by gene or domain duplication and sequence divergence, and type II AFPs represent a prominent example of parallel evolution. Despite their distinct origins and structures, all antifreeze proteins inhibit ice growth in similar ways — a remarkable display of functional convergence.

The ancestries of cod AFGP, type I AFP of pleuronectiformes (righteye flounders and plaice)[72–74] and cottidae (short-horn and grubby sculpins),[75–76] and type III AFP of Antarctic eel pouts[77,78] and their counterparts Northern Hemisphere[79,80] remain to be deciphered. Type I AFP of the unrelated flat fishes and cottids (Fig. 1) are very similar in being Ala-rich (>60%), and α-helical in structure,[81] and presumed homologous. The flatfish AFPs have a prominent repetitive sequence — three to five tandem repeats of an 11-residue unit, Thr-X_2-(Asp/Asn) X_7 (where X is mostly Ala), suggesting tandem duplications had occurred in the evolution of their encoding genes. By contrast, sculpin AFP has much less regularity in the 11-residue repeat structure, and differ from flounder AFP in N-terminal sequence, in having a blocked N-terminus[82] and a different ice crystallographic plane of adsorption.[74] Thus, whether flatfish and sculpin AFPs are indeed homologous, or are derived through convergent evolution from two look alike ancestral sequences remain to be verified. An additional intrigue in the evolution of type I AFP is the presence of a non-liver, intracellular subtype in the winter flounder, first discovered in skin and later in other peripheral tissues.[82] This skin AFP subtype is apparently encoded by a family of genes distinct from the gene family that encodes the secreted liver-type. It would be of great interest to determine how these two gene families evolved within the same fish.

The cooling of the polar and subpolar oceans had created a "hot bed" of antifreeze evolution. AF(G)Ps with their ice-binding function, arguably, are truly novel proteins. Deciphering the origins of the diverse fish antifreezes and the molecular processes by which they arose will richly broaden our knowledge of how novel protein gene and function arose, and advance the field of molecular evolution in general.

References

1. Scholander PF, van Dam L, Kanwisher JW, Hammel HT and Gordon MS (1957). Supercooling and osmoregulation in Arctic fish. *J. Cell Comp. Physiol.* **49**: 5–24.
2. Leim AH and Scott WB (1966). *Fishes of the Atlantic Coast of Canada*. Fish Res Board Can, Ottawa, p. 357.
3. Raymond JA (1993). Glycerol and water balance in a near-isosmotic teleost, winter-acclimatized rainbow smelt. *Can. J. Zool.* **71**: 1849–1854.
4. Kennett JP (1977). Cenozoic evolution of Antarctic glaciation, the circum-Antarctic Ocean, and their impact on global paleocenography. *J. Geophys. Res.* **82**: 3843–3860.
5. Kennett JP (1982). *Marine Geology*. Prentice-Hall, New Jersey.
6. Cheng C-HC (1998). Evolution of the diverse antifreeze proteins. *Curr. Opin. Genet. Dev.* **8**: 715–720.
7. Davies PL and Sykes BD (1997). Antifreeze proteins. *Curr. Opin. Struct. Biol.* 7: 828–834.
8. Raymond JA and DeVries AL (1977). Adsorption inhibition as a mechanism of freezing resistance in polar fishes. *Proc. Natl. Acad. Sci. USA* **74**: 2589–2593.
9. Hoshino T, Odaira M, Yoshida M and Tsuda S (1999). Physiological and biochemical significance of antifreeze substances in plants. *J. Plant Res.* **112**: 255–261.
10. Hon WC, Griffith M, Mlynarz A, Kwok YC and Yang DS (1995). Antifreeze proteins in winter rye are similar to pathogenesis-related proteins. *Plant Physiol.* **109**: 879–889.
11. Andorfer CA and Duman JG (2000). Isolation and characterization of cDNA clones encoding antifreeze proteins of the pyrochroid beetle *Dendroides canadensis. J. Insect Physiol.* **46**: 365–372.
12. Liou YC, Thibault P, Walker VK, Davies PL and Graham LA (1999). A complex family of highly heterogeneous and internally repetitive hyperactive antifreeze proteins from the beetle *Tenebrio molitor. Biochemistry* **38**: 11415–11424.
13. Nelson NS (1994). *Fishes of the World*. John Wiley and Sons, New York.
14. Ewart KV and Fletcher GL (1990). Isolation and characterization of antifreeze proteins from smelt (*Osmerus mordax*) and Atlantic herring (*Clupea harengus harengus*). *Can. J. Zool.* **68**: 1652–1658.

15. Ewart KV and Fletcher GL (1993). Herring antifreeze protein: primary structure and evidence for a C-type lectin evolutionary origin. *Mol. Mar. Biol. Biotech.* **2**: 20–27.
16. Ng NF, Trinh YK and Hew CL (1986). Structure of an antifreeze polypeptide precursor from the sea raven, *Hemitripterus americanus. J. Biol. Chem.* **261**: 15690–15696.
17. Ewart KV, Rubinsky B and Fletcher GL (1992). Structural and functional similarity between fish antifreeze proteins and calcium-dependent lectins. *Biochem. Biophys. Res. Commun.* **185**: 335–340.
18. Drickamer K (1999). C-type lectin-like domains. *Curr. Opin. Struct. Biol.* **9**: 585–590.
19. Ewart KV, Li Z, Yang DSC, Fletcher GL and Hew CL (1998). The ice-binding site of Atlantic herring antifreeze protein corresponds to the carbohydrate-binding site of C-type lectins. *Biochemistry* **37**: 4080–4085.
20. Gronwald W, Loewen MC, Lix B, Daugulis A, Sönnichsen F, Davies PL and Sykes BD (1998). The solution structure of type II antifreeze protein reveals a new member of the lectin family. *Biochemistry* **37**: 4712–4721.
21. Deng G, Andrews DW and Laursen RA (1997). Amino acid sequence of a new type of antifreeze protein, form the longhorn sculpin *Myoxocephalus octodecimspinosis. FEBS Lett.* **402**: 17–20.
22. Li WH, Tanimura M, Luo CC, Datta S and Chan L (1988). The apolipoprotein multigene family: biosynthesis, structure, structure-function relationships, and evolution. *J. Lipid Res.* **29**: 245–271.
23. Baker WC and Dayhoff MO (1977). Evolution of lipoproteins deduced from protein sequence data. *Comp. Biochem. Physiol.* **576**: 309–315.
24. Fitch WM (1977). Phylogenetics constrained by the crossover process as illustrated by human hemoglobins and a thirteen cycle, eleven amino acid repeat in human apolipoprotein A-I. *Genetics* **86**: 623–644.
25. Borhani W, Rogers DP, Engler JA and Brouillette CG (1997). Crystal structure of truncated human apolipoprotein A-I suggests a lipid-bound conformation. *Proc. Natl. Acad. Sci. USA* **94**: 12291–12296.
26. Wilson C, Wardell MR, Weisgraber KH, Mahley RW and Agard DA (1991). Three-dimensional structure of the LDL receptor binding domain of human apolipoprotein E. *Science* **252**: 1817–1822.
27. Ohta T (1989). Role of gene duplication in evolution. *Genome* **31**: 304–310.

28. Graur D and Li W-H (2000). *Fundamentals of Molecular Evolution.* Sinauer, Massachusetts, pp. 249–322.
29. DeVries AL (1971). Glycoproteins as biological antifreeze agents in Antarctic fishes. *Science* **172**: 1152–1155.
30. DeVries AL, Vandenheede J and Feeney RE (1971). Primary structure of freezing point-depressing glycoproteins. *J. Biol. Chem.* **246**: 305–308.
31. DeVries AL (1982). Biological antifreeze agents in coldwater fishes. *Comp. Biochem. Physiol. A* **73**: 627–640.
32. Cheng C-HC (1996). Genomic basis for antifreeze glycopeptide heterogeneity and abundance in Antarctic fishes. In: Ennion S, Goldspink G (eds.), *Gene Expression and Manipulation in Aquatic Organisms.* Cambridge, United Kingdom, pp. 1–20.
33. Morris HR, Thompson MR, Osuga DT, Ahmed AT, Chan SM, Vandenheede JR and Feeney RF (1978). Antifreeze glycoproteins from the blood of an Antarctic fish. *J. Biol. Chem.* **253**: 5155–5162.
34. Cheng C-HC and Chen L (1999). Evolution of an antifreeze glycoprotein. *Nature* **401**: 443–444.
35. Hsiao KC, Cheng C-HC, Fernandes IE, Detrich HW and DeVries AL (1990). An antifreeze glycopeptide gene from the Antarctic cod *Notothenia coriiceps neglecta* encodes a polyprotein of high peptide copy number. *Proc. Natl. Acad. Sci. USA* **87**: 9265–9269.
36. Chen L, DeVries AL and Cheng C-HC (1997a). Evolution of antifreeze glycoprotein gene from a trypsinogen gene in Antarctic notothenioid fish. *Proc. Natl. Acad. Sci. USA* **94**: 3811–3816.
37. Chen L, DeVries AL and Cheng C-HC (1997b). Convergent evolution of antifreeze glycoproteins in Antarctic notothenioid fish and Arctic cod. *Proc. Natl. Acad. Sci. USA* **94**: 3817–3822.
38. Leaver MJ and George SG (1996) unpublished. Genbank accession number X56744.
39. Levinson G and Gutman GA (1987). Slipped-strand mispairing: a major mechanism for DNA sequence evolution. *Mol. Biol. Evol.* **4**: 203–221.
40. Cheng C-HC (1998). Origin and mechanism of evolution of antifreeze glycoproteins in polar fishes. In: di Prisco G and Pisano E (eds.), *Fishes of Antarctica. A Biological Overview.* Springer-Verlag, Italia, pp. 311–328.
41. Li W-H, Luo C-C and Wu C-I (1985). Evolution of DNA sequences. In: MacIntyre RJ (ed.), *Molecular Evolutionary Genetics.* Plenum, New York, pp. 1–94.

42. O'Grady SM, Schrag JD, Raymond JA and DeVries AL (1982). Comparison of antifreeze glycopeptides from Arctic and Antarctic fishes. *J. Expt. Zool.* **224**: 177–185.
43. Fletcher GL, Hew CL and Joshi SB (1982). Isolation and characterization of antifreeze glycoproteins from the frostfish, *Microgadus tomcod. Can. J. Zool.* **60**: 348–355.
44. Svetovidov AN (1948). *Gadiformes.* Israel program for scientific translation, Jerusalem.
45. Cohen DM, Inada T, Iwamoto T and Scialabba N (1990). FAO fisheries synopsis No. 125, Vol. 10: *Gadiform Fishes of the World.* United Nations, Rome.
46. Howe GJ (1991). Biogeography of gadoid fishes. *J. Biogeogr.* **18**: 595–622.
47. Clarke A and Johnston IA (1996). Evolution and adaptive radiation of Antarctic fishes. *Trends Ecol. Evol.* **11**: 187–228.
48. Shackleton NJ, Backman J, Zimmerman J, Kent DV, Hall MA, Roberts DG, Schnitker D, Baldauf JG, Desprairies A, Homrighausen R, Huddlestun P, Keene JB, Kaltenback AJ, Krumsiek KAO, Morton AC, Murray JW and Westberg-Smith J (1984). Oxygen isotope calibration of the onset of ice-rafting and history of glaciation in the North Atlantic region. *Nature* **307**: 620–623.
49. Xu M and Lewis RW (1990). Structure of a Protein Superfiber: Spider Dragline Silk. *Proc. Natl. Acad. Sci. USA* **87**: 7120–7124.
50. Tsitilou SG, Rodakis GC, Alexopoulou M, Kafatos FC, Ito K and Iatrou K (1983). Structural features of β family chorion sequences in the silkmoth *Bombyx mori*, and their evolutionary implications. *EMBO J.* **2**: 1845–1852.
51. Pau RN (1984). Cloning of cDNA for a juvenile hormone-regulated oothecin mRNA. *Biochim. Biophys. Acta* **782**: 422–428.
52. Robson P, Wrights GN, Sitarz E, Mait A, Rawat M, Youson JH and Keeley FW (1993). Characterization of lamprin, an unusual matric protein from lamprin cartilage — Implications for evolution, structure, and assembly of elastin and other fibrillar proteins. *J. Biol. Chem.* **268**: 1440–1447.
53. Doolittle RF (1994). Convergent evolution: the need to be explicit. *Trends Biochem. Sci.* **19**: 15–18.
54. Cvekl A and Piatigorsky J (1996). Lens development and crystallin gene expression: many roles for Pax-6. *Bioessays* **18**: 621–630.

55. Yokoyama S (2000). Molecular evolution of vertebrate visual pigments. *Prog. Retin. Eye Res.* **19**: 385–419.
56. Tan Y and Li WH (1999). Trichromatic vision in prosimians. *Nature* **402**: 36.
57. Eastman JT and Clarke A (1998). A comparison of adaptive radiations of Antarctic fish with those of non-Antarctic fish. In: di Prisco G and Pisano E (eds.), *Fishes of Antarctica. A Biological Overview*. Springer-Verlag, Italia, pp. 1–26.
58. Lawver LA, Gahagan LM and Coffin MF (1992). The development of paleoseaways around Antarctica. In: Kennett JP and Warnke DA (eds.), *The Antarctic Paleoenvironment: A Perspective on Global Change. Part One. Antarctic Research Series*, Vol. 56. American Geophysical Union, Washington DC, pp. 7–30.
59. Foster TD (1984). The marine environment. In: Laws RM (ed.), *Antarctic Ecology*, Vol. 2. Academic Press, London, pp. 345–371.
60. Grande L and Eastman JT (1986). A review of Antarctic ichthyofaunas in the light of new fossil discoveries. *Paleontol.* **29**: 113–137.
61. Grande L and Chatterjee S (1987). New Cretaceous fish fossils from Seymour Island, Antarctic Peninsula. *Paleontol.* **30**: 829–837.
62. Long DJ (1992). Sharks from the La Meseta Formation (Eocene), Seymour Island, Antarctic Peninsula. *J. Vert. Paleontol.* **12**: 11–32.
63. Bargelloni L, Ritchie PA, PatarnelloT, Battaglia B, Lambert DM and Meyer A (1994). Molecular evolution at subzero temperatures: mitochondrial and nuclear phylogenies of fishes from Antarctica (*Suborder notothenioidei*) and the evolution of antifreeze glycopeptides. *Mol. Biol. Evol.* **11**: 854–863.
64. Martin AP and Palumbi S (1993). Body size, metabolic rate, generation time, and the molecular clock. *Proc. Nat'l. Acad. Sci. USA* **90**: 4087–4091.
65. Brown WM, George MG and Wilson AC (1979). Rapid evolution of animal mitochondrial DNA. *Proc. Natl. Acad. Sci. USA* **76**: 1976–1971.
66. DeVries AL and Lin Y (1977). Structure of a peptide antifreeze and mechanism of adsorption to ice. *Biochim. Biophys. Acta* **495**: 388–392.
67. Scott GK, Davies PL, Shears MA and Fletcher GL (1987). Structural variations in the alanine-rich antifreeze proteins of the Pleuronectinae. *Eur. J. Biochem.* **168**: 629–633.

68. Knight CA, Cheng C-HC and DeVries AL (1991). Adsorption of α-helical antifreeze peptides on specific ice crystal surface planes. *Biophys. J.* **59**: 409–418.
69. Hew CL, Joshi S, Wang N-C, Cao M-H and Ananthanarayanan VS (1985). Structures of shorthorn sculpin antifreeze polypeptides. *Eur. J. Biochem.* **151**: 167–172.
70. Chakrabartty A, Hew CL, Shears M and Fletcher GL (1988). Thr primary structures of the alanine-rich antifreeze peptides from gruppy sculpin, *Myoxocephalus aenaeus. Can. J. Zool.* **66**: 403–408.
71. Cheng C-HC and DeVries AL (1989). Structures of antifreeze peptides from the antarctic eel pout, *Austrolycichthys brachycephalus. Biochim. Biophys. Acta* **997**: 55–64.
72. Wang X, DeVries AL and Cheng C-HC (1995). Antifreeze peptide heterogeneity in an Antarctic eel pout includes an unusually large major variant comprised of two 7 kDa type III AFPs linked in tandem. *Biochim. Biophys. Acta* **1247**: 163–172.
73. Hew CL, Wang N-C, Joshi S, Fletcher GL, Scott GK, Hayes PH, Buettner B and Davies PL (1988). Multiple genes provide the basis for antifreeze protein diversity and dosage in the ocean pout, *Macrozoarces americanus. J. Biol. Chem.* **263**: 12049–12055.
74. Scott GK, Hayes PH, Fletcher GL and Davies PL (1988). Wolffish antifreeze protein genes are primarily organized as tandem repeats that each contain two genes in inverted orientation. *Mol. Cell. Biol.* **8**: 3670–3675.
75. Yang DS, Sax M, Chakrabartty A and Hew CL (1988). Crystal structure of an antifreeze polypeptide and its mechanistic implications. *Nature* **333**: 232–237.
76. Gong Z, Ewart KV, Hu Z, Fletcher GL and Hew CL (1996). Skin antifreeze protein genes of the winter flounder, *Pleuronectes americanus*, encode distinct and active polypeptides without the secretory signal and prosequences. *J. Biol. Chem.* **271**: 4106–4112.

Chapter 5

The Structure of Fish Antifreeze Proteins

Darin J Brown and Frank D Sönnichsen
Department of Physiology and Biophysics
Case Western Reserve University
Cleveland, OH, USA

Introduction

Since the discovery of antifreeze proteins and glycoproteins (AF(G)Ps) three decades ago[1] yielded the explanation for survival of teleost fish at temperatures below the colligative freezing point,[2] many researchers have sought to determine the distinct function and nature of these exciting proteins. AF(G)Ps possess the unique ability to recognize and bind to growing ice crystals, thus depressing the freezing point in a non colligative manner and protecting these polar fish from fatally freezing in their ice-laden environments.[3,4] These proteins further inhibit ice recrystallization, reducing shear forces and other damaging effects during freeze-thaw cycles.[5–7] Some also have shown to stabilize biological membranes in a presumably unrelated activity,[8] providing additional protective benefits for the organism and cells in freezing environments. Indeed, a surprisingly large number of proteins of different origin and structure have been shown to exhibit these functions. At the present time, five different types of AF(G)Ps have been identified from the plasma sera of several species of teleost fish, and tissue-specific isoforms are emerging in increasing number.[9–11] The last decade has seen a surge in our knowledge of AFP structure. Today, the high resolution structures for all but the most recently discovered fish AF(G)Ps have been determined, as well as structures for the first insect AFPs have also been solved.[12,13] Analyses of these structures have facilitated further studies on molecular elements and interactions involved in ice recognition by AF(G)Ps, and provided many insights into the mechanism of ice recognition and adsorption. These findings include the identification of particular residues and surfaces critical for activity by biochemical analyses, while biophysical studies have shed light on the intriguing process of AF(G)P specific binding to ice lattice planes.

All AF(G)Ps appear to function similarly in their basic underlying activity of binding ice surfaces, but they exhibit a remarkable divergence in their structural and dynamic properties and in specific details of their mechanism. However, the diversity inherently offered the opportunity to identify the essential structural properties, and it is the focus of this chapter to review the structural progress, highlight the differences and commonalties in fish AFP properties, and discuss the mechanistic implications. Yet, in the following it will become quickly apparent that the tremendous progress in our biochemical and biophysical knowledge to this point has not resulted in the unambiguous identification of common, critical structural motifs or the emergence of a mechanistic and thermodynamic understanding of AF(G)P function.

Antifreeze Glycoproteins

Initially discovered as the first biological antifreeze molecules,[15] antifreeze glycoproteins (AFGPs) are not only a distinct group, they are different in several aspects from the AFPs (Table 1). Their peptide backbone structure is extremely simple, commonly consisting of Ala-Ala-Thr tripeptide repeats. Each of the Thr residues is O-glycosylated by a disaccharide (N-acetyl-D-galactosamine (α1-3) galactose) thought to be extended from the polypeptide backbone (Fig. 1). Differences in the number of repeats (between four and 30 units) and also the occasional incorporation of proline residues leads to a significant degree of variability between AFGP isoforms despite the protein's simplicity. Eight different fractions have been characterized with molecular masses ranging from 2.7 to 32 kDa.[1]

Structurally, AFGPs have been the most elusive proteins with antifreeze capability to characterize. Several studies have attempted to determine the backbone structure of these polypeptides and to establish

Table 1 Physical properties of fish AF(G)Ps.*

Properties	AFGP	AFP I	AFP II	AFP III	AFP IV
Species	Cod and Antarctic notothenioid	Winter flounder, Alaskan plaice, and Arctic sculspin	Herring, smelt, and sea raven	Eel pout, ocean pout, and wolffish	Longhorn sculpin
Heterogeneity	Polymeric lengths	Subtypes: repetitive/ non-repetitive, and tissue-specific	Subtypes: Ca^{2+} independent/ dependent	Isoforms Intramolecular dimer	NA
M	2.7–32 kDa	3–5 kDa	14–24 kDa	7 kDa	12.3 kDa
Amino acid bias	>60% Ala, >30% Thr	>60%	8% Cys	NA	17% Gln
Secondary structure	Polyproline II, helices and γ-turns	100% α-helix at 0°C	Mixed; Two helices and two antiparallel β-sheets from nine β-strands	Two triple-stranded β-sheets and single helical turn	Mostly α-helix
Tertiary structure	Extended, flexible coil	Single amphipathic α-helix with salt bridge	Globular, CRD fold of C-type lectins with five disulfide bonds	Globular, β-sandwich fold with flat surfaces	Antiparallel four-helix bundle
Ice-binding plane	Close to $1\,0\,\bar{1}\,0$	$2\,0\,\bar{2}\,1/2\,\bar{1}\,\bar{1}\,0$	$1\,1\,\bar{2}\,1$/NA	$1\,0\,\bar{1}\,0, + 2\,0\,\bar{2}\,1$ or others concurrently	NA
Ice-binding surface properties	H-bonding via axial hydroxyl groups of sugar units	Two putative sites: hydrophilic TAN face or hydrophobic TAA face	Planar surface of Ca^{2+}-binding loop and hydrophilic residues + hydroxyl groups of residues in extension	Five polar groups embedded in nonpolar planar surface	NA

*Adapted from Davies and Sykes.[14]

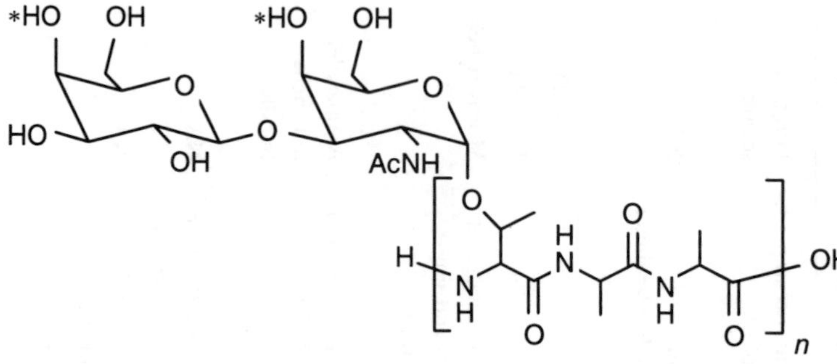

Fig. 1 AFGP repeat unit of dialanine and O-glycosylated threonine residues. Axial hydroxyls putatively involved in ice-binding are marked with *.

the spatial relationship of the sugar moiety with respect to the backbone. Over the years, a number of structural models have been proposed.[16] Among them, the presence of γ-turns based on FT-IR experiments[17] and polyproline type II helices have been suggested,[18] while α-helical backbone conformations were excluded.[19,20] Most recently, the structure of AFGP-8 and AFGP-1 to -5 have been re-investigated using NMR-spectroscopy.[21,22] These studies concluded that even at temperatures near the freezing point, neither strongly preferred backbone conformations nor preferred sugar-peptide orientations could be detected. AFGP-1 to -5 have no long-range order, and all observations are consistent with a highly flexible, random-coil structure.[22] Although largely similar in properties, the degree of backbone flexibility in AFGP-8 is lower, resulting in a somewhat more rigid conformation. The proline residues in this small peptide lead to slightly incomplete conformational averaging, consistent with a more extended, polyproline type II helical fold. Overall, the characterized peptides are comprised of several repeating units, and have shown absence of long or medium range order. Therefore, it seems unlikely that further structural stabilization will occur in larger AFGP on the basis of increased repeat numbers despite the increased opportunity of inter-repeat interactions. More realistically, in contrast to the conformationally stable AFPs, AFGPs of all sizes adopt a variety of conformations, among which a preferred ice-binding conformation exists and is selected upon ice adsorption. AFGPs may then serve as an example that a defined stable tertiary structure is not prerequisite for antifreeze activity, even though their activity is relatively lower than that of AFPs.

In agreement with traditional mechanistic viewpoints, the disaccharide moieties in AFGPs are believed to participate in hydrogen bonding to the ice surface. The regular spacing of the disaccharide units on one side of the peptide facilitated by regular backbone conformations such as the polyproline helical folds[18] is envisioned to lead to multiple hydrogen bond formation by the axial hydroxyl groups of the disaccharides and peptide adsorption to the $(1\ 0\ \bar{1}\ 0)$ prism plane of ice.[23] Direct evidence for the involvement of the sugar units in ice adsorption has been derived from the inhibition of AFGP activity by boronic acid and its derivatives. Boronic acid has been shown to bind the axial hydroxyl groups extending from the disaccharide unit causing a loss in activity and consequently inhibiting ice-binding.[24,25] Though, it appeared questionable that two hydrogen bonds per sugar unit would be sufficient for irreversible ice-binding of a small AFGP, such as AFGP-8 that contains four sugar units. An elegant explanation was forwarded by Knight and DeVries[23]; they suggested the partial incorporation of the axial hydroxyl groups into the ice lattice stabilized by three hydrogen bonds per hydroxyl group, thus resulting in 24 hydrogen bonds per peptide. Further studies on the molecular features critical for AFPG activity have been limited due to the restrictions of the natural protein source. In contrast to other AFPs, bacterial expression is not feasible, and chemical synthesis, despite promising beginnings, has yet to provide sufficient quantities for structure-function studies. Nevertheless, one can expect significant improvements in our understanding of the function of AFGPs, once structural AFGP features can be systematically analyzed using chemically synthesized analogs.[26–28]

Antifreeze Proteins

AFPs overall display generally well-defined folds and conformations in contrast to the AFGPs, yet show few common properties and specific features, except for their antifreeze activity and ability to bind to ice surfaces. The structures of type I, II and III AFPs, all of which have been determined by X-ray crystallography and/or NMR spectroscopy are as diverse as their primary structures, which will become more evident in the following discussion of their secondary or tertiary structural features.

Among the known AFPs, only the recently identified and cloned type IV AFP from the longhorn sculpin has yet to have its three-dimensional structure determined. The initial characterization of this 108-amino acid protein by circular dichroism studies confirmed a folded protein comprised of a significant amount of α-helical regions.[29] The primary sequence of this protein contains about 20% similarity to several apolipoproteins, which led to predictions of this protein comprising a lipoprotein-like four-helical bundle fold (Fig. 2).[14,29] Hydrophobic residues, which would be localized within the core and are essential for stability of the helical-bundle fold, are particularly conserved.

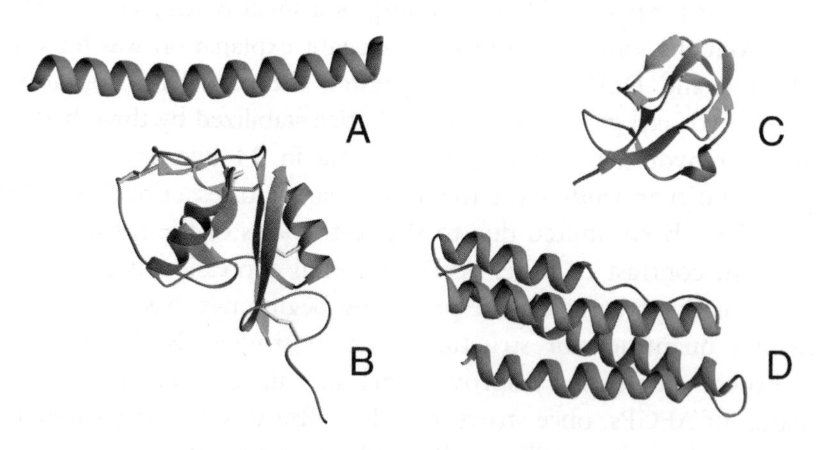

Fig. 2 Ribbon models of the three-dimensional structures of fish antifreeze proteins. **(A)** Type I α-helix {Sicheri & Yang (1995) #10500} (PDB accession code 1WFA), **(B)** Type II C-type lectin CRD-like globular fold {Gronwald, Loewen *et al.* (1998) #3900} (PDB 2AFP), **(C)** Type III globular β-sandwich fold {Sönnichsen, DeLuca *et al.* (1996) #4170} {Jia, DeLuca *et al.* (1996) #10570} (PDB 1KDF & 1MSI), and **(D)** Model of type IV antiparallel helical fold {Zhao, Deng *et al.* (1998) #6870} (based on the structure for apolipoprotein III) {Breiter Holden Biochemistry (1991)} (PDB 1AEP).

Recombinant type IV displays thermal hysteresis activity and yields crystals with hexagonal trapezohedral shape within the thermal hysteresis window, both properties establishing it as an AFP. The upper and lower halves of this crystal are rotated contrary to each other by 20°, thus displaying this unique shape rather than classical hexagonal bipyramidal crystal morphology. If interpreted in terms of binding planes, the macroscopic crystal shape reflects protein-binding to the ($3\ \bar{1}\ \bar{2}\ 1$) plane; a plane uncommon among AFPs.[30] Although the crystal dimensions (c/a-axis ratio = 3.2:1) are similar to those of flounder type I AFP, the type IV ice-etching pattern is inconsistent with the flounder adsorption plane. It shows an unusual sausage-shaped pattern that may imply a possible association of the protein with multiple binding planes or reduced binding specificity. Overall, it will be interesting to see the location and nature of the ice-binding surface along with the backbone fold confirmed by a detailed structural analysis.

Type I AFP

As the most extensively studied antifreeze proteins, type I AFPs have provided a wealth of data regarding structure, ice-binding, and ice recognition, and have served as a model system for the development and critical evaluation of mechanistic ideas. Foremost, type I AFPs exist abundantly in several different fish species and isoforms as 3–5 kDa monomers (Table 1). X-ray crystallography yielded the first high-resolution view of the type I AFP structure, that of a 37-amino acid liver-specific winter flounder isoform (HPLC6). The protein is folded into a slightly curved amphipathic, single α-helical fold that defines all type I AFP isoforms[31] (Fig. 2). The helical structure over the entire sequence length can be attributed to the large proportion of alanine residues and N- and C-terminal capping sequences that reduce both end-effects and a destabilizing interaction between helix dipole and N-terminal charge. The shape of this fold facilitated identification of the stereo-specific adsorption direction from ice-etching studies.[32] The elliptical etching pattern indicated protein adsorption occurred in the ⟨$0\ 1\ \bar{1}\ 2$⟩ direction of ice,[33] while the location of the pattern specified the (2 0 $\bar{2}$ 1) pyramidal plane as the adsorption plane. Small ice crystals, grown in the hysteresis assay of HPLC6, displayed a hexagonal bipyramidal morphology with a defined c/a axis ratio of 3.3:1.[34] This ratio suggests that for this protein, the macroscopically expressed crystal surface corresponds with the adsorption plane determined by ice etching.

HPLC6 along with other winter flounder and plaice isoforms exhibits a primary sequence that can be expressed as 11-residue repeats. A slightly smaller than ideal helical pitch results in a linear array of Thr and Asp/Asn residues on one side of the helix, while the remaining sides consist of mainly Ala-side chains, one salt bridge and the capping sequences. The separation between equivalent residues in the repeats is 16.5 Å along the helical axis, which matches the 16.7 Å periodicity of water molecules in the ⟨0 1 $\bar{1}$ 2⟩ adsorption direction of the ice-binding plane. This periodicity match along with a number of experimental observations from residue replacement studies[35] led to suggestions that type I AFPs bind to ice via hydrogen bonding. The complementary spacing between the Thr and Asn side chains and the water molecules in the crystal lattice facilitates the formation of multiple H-bonds are perceived to provide the source of the AFP-binding affinity and specificity. Two separately derived models using these concepts differ only in the positioning of the AFP on the ice surface and the conformation of ice-binding residues.[31,36] In both cases, the preferred side chain conformations might appear at odds with the solution characterization of HPLC6 by NMR spectroscopy.[37] This study showed the absence of strong rotamer preferences for the Thr residues even at temperatures near the freezing point. However, the small entropic cost in ordering these side chains upon ice adsorption is expected to be easily overcome by favorable interactions between protein and ice.

Nevertheless, further concerns about the significance of hydrogen bonds initiated a number of studies in different laboratories that have questioned the role of Thr residues involved in ice-binding. The replacement of initially the two central and later all Thr residues in HPLC6 by isostructural Val residues yielded an analog with slightly reduced activity, while the replacement with Ser residues, a change that retains the hydrogen bonding capability of the side chain, abolished antifreeze activity.[38–40] While this observation appears to attribute a significant role to the Thr methyl group, the hydrophobic contribution alone does not support efficient thermal hysteresis activity.[41] These results demonstrate that steric integrity of the protein surface is crucial and hydrogen bonding alone remains insufficient to characterize the role of the Thr residues in this AFP.

Subsequent studies further suggest that replacements of Asn/Asp residues in the hydrophilic face, and of Leu residues, a third residue in the proposed ice-binding motif,[31] are not necessarily indicative of their involvement in ice-binding, rather that these residues are involved in preventing protein aggregation in solution.[42] The individual replacement of four Ala residues on each side of the helix by leucines, that utilize the steric sensitivity of AFP-ice binding, yielded two analogs (A17L and A20L) with significantly reduced antifreeze activities.[43] This elegant study suggests that the hydrophilic face of type I AFP might not be the ice-binding site, but that the protein adsorbs to ice utilizing another surface comprised of Thr and Ala residues. This alternative surface, rotated by about 90° with respect to the hydrophilic face,[44] is still relatively planar. This putative, hydrophobic binding is consistent with the activity of all but one type I AFP analogs, a double Gln substitution for Asn residues lying outside of either proposed ice-binding surface that abolished all activity.[42] The site is also largely consistent with studies on skin-type isoforms,[9] and synthetic analogs aimed at testing the role of lysine residues in ice-binding.[45,46] Moreover, comparing all known type I isoforms without bias towards particular residues, this protein surface (TAA) was suggested to be highly conserved.[43] Thus, for the first time an ice-binding site was proposed that would be largely common among all type I AFPs including the isoforms identified in shorthorn and grubby sculpins. These isoforms largely lack Thr and Asn residues and 11 amino acid repeats. Based on the presumed H-bonding requirement, the source of their ice-binding activity had been previously ascribed to several Lys-residues.[47] The realignment of sequences suggest the possibility that these isoforms bind ice similarly through an all alanine face, identical to that of HPLC6 with the exception of the Thr residues. Supportive evidence for this conclusion can be drawn from ice-etching studies. In contrast to HPLC6, the isoform SS-8 from sculpin adsorbs specifically to the secondary prism planes of ice,[33] and the same binding specificity has been observed for an analog of HPLC6, in which all Thr are replaced by Ala residues.[44] The implications of this observation are far reaching for the discussion of the AFP-binding specificity, if we accept the suggestion that the Ala-rich face is the ice-binding site for both the SS-8 and HPLC6 isoforms: first, hydrogen bonds are required neither for the binding affinity nor for the binding specificity. Second, the data suggest that the hydrophobic alanine-face has inherently high affinity to both the secondary prism plane and the $(2\ 0\ \bar{2}\ 1)$ plane. Thirdly, the question is raised whether the specificity demonstrated in ice-etching studies that are performed at an optimized, low protein concentration of the type I AFP is overestimated, and the protein binds two (or more?) planes at physiological concentrations? Certainly, the observation of needle-like crystals for type I AFP, and the observation of two adsorption planes for insect THPs would provide credence for multiple binding planes.

In conclusion, despite the apparent simplicity of type I AFP on a sequence and structural level, and the large number of studies that focused on this protein due to its "model" character and made it the best characterized AFP, we are still far away from understanding the basic properties of this protein. Further studies will be needed to unambiguously establish the actual binding surface, and the kinetic and energetic details of the ice adsorption.

Type II AFP

Type II AFPs exist in three fish species: sea raven, smelt and Atlantic herring,[48,49] and all three isoforms contain a 30% sequence homology with the carbohydrate-recognition domain (CRD) of Ca^{2+}-dependent (C-type) lectins. C-type lectins are a widespread family of receptor and proteins involved in cell adhesion, surface recognition and extracellular matrix formation in which the CRD domain binds sugars specifically through contact with calcium. The extent of sequence similarity between lectin CRD domains and the type II AFPs is confined to short sequences and single amino acids at intervals throughout the protein corresponding to essential residues, disulfide-bonded cysteines, and Ca^{2+}-binding motifs found within the lectin sequence.[48,50] Initial modeling studies suggested that the type II AFP sequences are highly compatible with the CRD fold of C-type lectins.[50,51] Experimentally, NMR characterization of the ^{15}N-labeled sea raven type II AFP confirmed the fold.[52] The three-dimensional structure displays a single globular domain consisting of two helices and nine β-strands in two β-sheets stabilized by five disulfide bonds, two of which are unique to the AFP fold.

Both the herring and smelt type II AFPs demonstrate Ca^{2+}-dependent antifreeze activity, and on the sequence level, contain semi-conserved residues in the Ca^{2+}/carbohydrate-binding site of the CRD. Mutating a loop comprising the coordinating QPD sequence of the herring type II AFP resulted in loss of both thermal hysteresis activity and ice crystal modification capability. This suggested the carbohydrate binding site is part of the ice-binding surface in agreement with the implications of the homology, and the surface planarity in this region of the AFP models.[51,53] This putative ice-binding site is created by one β-sheet and a spatially adjacent loop region, and is rich in hydrophilic residues.

Although largely alike, the herring and smelt isoforms with 40% sequence similarity, sea raven type II AFP lacks the Ca^{2+}-binding ligating side chains in the carbohydrate-binding site, and expectedly exhibits high-level, Ca^{2+}-independent antifreeze activity. Nevertheless, based on the sequence homology and the surface planarity in the model, the protein surface that comprises the Ca^{2+}/carbohydrate-binding site in the lectin was suggested to be the ice-binding site of the sea raven isoform.[50] This hydrophilic surface was shown to provide a reasonable fit to the $(1\ 1\ \bar{2}\ 1)$ bipyramidal plane, the established adsorption plane,[54] by forming integral hydrogen bonds with the ice lattice in modeling studies.[55] Surprisingly, extensive mutational analysis of the sea raven ice-binding site yielded fairly different results.[56] The mutation of many hydrophilic residues in this presumed ice-binding surface, more precisely near the β-sheet stabilizing the carbohydrate-binding site, did not affect antifreeze activity. Even significant changes in the encompassed carbohydrate-binding loop itself reduced activity to a much smaller degree than expected or seen with the herring type II isoform. The identification of Ser-120 present at the edge of the surface as a pivotal ice-binding residue, might suggest that a different protein surface is utilized, one that is more closely related to the active site of the homologous pancreatic stone protein.[57] These observations leave the nature of the sea raven AFP ice-binding site uncertain and further studies will be needed outline the location and exact nature of the sea raven-binding site.

Type III AFP

Identified in Arctic and Antarctic zoarcid fish (eel or ocean pouts), the primary sequence of the type III AFP is comprised of about 65 amino acids with no particular bias.[58–60] Numerous isoforms have been identified and grouped into two major forms based on their surface charges and separation by ion-exchange chromatography.[59] Highly homologous, these exhibit sequence identities of >50% even between the groups. Following the initial characterization of the protein fold[61] as a compactly folded β-sandwich, a number of high-resolution structures of both the QAE[62–65] and SP-type isoforms[66] have been solved. While some differences in the secondary structure contents are primarily a consequence of short and non-canonical curved strands, two antiparallel triple-stranded sheets are packed with nearly orthogonal orientation (Fig. 2). The fold has a single two-fold axis of symmetry and has been also described as a pretzel-fold. Low temperature factors determined by X-ray crystallography and NMR ^{15}N-relaxation studies in solution suggested a stable conformation for the entire protein backbone, and many of the surface residues were characterized as having high rotamer preferences at lower temperature.[62,67]

The protein exhibits fairly planar surfaces at mainly two sides. One of these surfaces, comprised of parts of the C-terminal sheet and spatially adjacent loops, has been unambiguously established as the protein's ice-binding surface. Five hydrophilic residues (Gln9, Asn14, Thr15, Thr18 and Gln44) in this surface have been mutated with consequential reduction of the thermal hysteresis activity.[68,69] Similarly, a number of nonpolar residues in or directly adjoining this surface reduce activity upon mutation into residues with different steric properties and further established the location of the active site.[70,71] In fact, very few mutations of the residues within this ice-binding surface led to uncompromised activity levels.[69] Notably replacements into larger residues are usually more detrimental than replacements into residues with smaller side chains. Accordingly, these studies suggest the importance of the entire surface rather than a select few residues contributing to the protein's activity. Also, protruding side chains are apparently not easily inserted in the crystal lattice, but seem to sterically inhibit proper interaction between ice and protein surface. Further, the relatively hydrophobic character of this surface was noted.[50] Using a neural network algorithm trained with 20 mutant type III proteins for which thermal hysteresis activity is known, Graether *et al.*[69] established the significance of this hydrophobicity in an unbiased fashion. In order to investigate the role of hydrogen bonds of the two Thr residues in this protein experimentally, the authors again pursued the individual Thr replacement by the isosteric nonpolar Val residue. Unfortunately, the results are not clear, as one analog could not be obtained while the reduced activity of the Thr15-Val analog, with 54% noticeably less active than the Thr15-Ser analog (70%), potentially originated from an unexpected local structural difference in the mutant AFP.

Ice lattice-bound models of type III AFP adsorbed to the $(1\ 0\ \bar{1}\ 0)$ prism plane of ice[54] have been proposed based on the specific spacing of polar residues.[62] In these models that differ by the protein orientation and the utilization of an extra row of water in the prism plane in one study,[63] numerous hydrogen bonds potentially form. Further, the sterically complementary protein and ice surfaces suggested significant intermolecular van der Waals interactions at the interface residues.[62] Particularly, it appeared that the protein polar side chains, embedded into the rather hydrophobic ice-binding surface with their hydrogen bonding groups barely surface accessible, are limited in their ability to form efficient bonds. Also noticeable was that a number of water molecules in the ice surface would not find a protein group for H-bonding in these models. Thus, the consequential net loss in H-bonds for protein and ice upon adsorption appeared incompatible with an H-bond dominated binding mechanism. The models did not sufficiently describe the source of the ice-binding specificity either, as it was noticed that the proposed lattice match was not unique, and other adsorption planes seemed feasible based on polar group spacing alone.[62] Interestingly, the re-evaluation of the protein's broad ice-etching pattern seems to suggest that the protein may bind to more than just the prism plane of ice, and this broad specificity was similarly seen in empirical energy calculations of the protein binding to models of five ice surfaces.[65] In this context, the solution structure of the RD3, an AFP consisting of two type III domains connected by a short nine-residue linker sequence (DGTTSPGLK), has also recently been solved and leads to an remarkable ice-bound model.[64] The linker places both domains in near ideal location to both simultaneously interact with the ice surface, which would explain the observed greater than two-fold activity increase for this type III intramolecular dimer. Either a small adjustment in linker conformation would allow the direct binding of both domains in identical fashion to a single ice surface or after an initial attachment of the C-terminal domain to the prism plane $\langle 0\ 0\ 0\ 1 \rangle$ direction, the N-terminal domain could bind in a subsequent step in the $\langle 0\ \bar{1}\ 0\ 1 \rangle$ direction to a plane somewhat recessed, i.e. behind the C-terminal domain adsorption plane. The latter binding mode would not require any relative domain rearrangements and would be consistent with the average structure of the calculated solution conformations, and again suggests the presence of more than one binding orientation.

To date, the ice-binding site of type III AFP is the best characterized ice-binding site among fish AFPs. It appears that the surface flatness and proper surface complementarity are more important for the activity of type III AFP than a specific spacing or periodicity of polar residues within this surface.[62,66] Still, a consensus model that describes the source of the ice affinity and specificity of this protein, as well as of AFPs in general, still needs to emerge. With the availability of efficient expression systems for several isoforms and its stable protein fold, type III AFP is nevertheless a prime candidate for further biophysical studies targeted at enhancing our understanding of the underlying principles of protein-ice adsorption.

Mechanism

The vast majority of experimental evidence supports the notion that AF(G)Ps act by directly binding to ice crystals under *in vitro* conditions. Although a focus of only one study, no effect of AFPs on bulk water properties has been observed[72] and little is known about AF(G)P hydration.[73,74] AF(G)P binding to ice, however, has been substantiated directly and indirectly by numerous studies. Most notably, freezing studies established at least partial AF(G)P incorporation into ice,[75] which contrasts the behavior of other aqueous solutes that concentrate in solution. This property is elegantly used in ice-etching studies pioneered by C. Knight[33] that provided the invaluable information on protein-binding specificity. Grain-boundary growth studies[76] and ellipsometry[77] further demonstrated AFGP-binding and yielded surface coverage estimates of 16% for the basal and 30% for prism surfaces. Recently, surface tunneling microscopy (STM) provided the first atomic level images of an ice surface in the presence of a type I AFP analog.[78] Surface indentations appropriate of the AFP dimensions support not only the type I AFP-binding orientation, but also the irregular adsorption of individual protein molecules. Moreover, within resolution limits, the average distances between AFPs resulted in an estimate of about 10% ice coverage.

The essential parts of the predominant adsorption-inhibition mechanism of action, proposed by Raymond and DeVries[75] are that AFPs adsorb to specific planes of the ice crystal surface and largely, but incompletely, cover these surfaces. A number of related mechanistic models, such as the mattress-button model or the step-pinning model utilize two- or three-dimensional Kelvin effects to explain growth inhibition for the ice front between surface bound AF(G)Ps.[33,36,79] The Kelvin or Gibbs-Thompson effect that relates an equilibrium phase transition point to the interface curvature is effectively demonstrated by the phenomenon of ice recrystallization. At limited warming rates, or temperatures below and near the freezing point, a partly frozen melt will reduce its free energy by recrystallization: larger crystals will grow at the expense of smaller crystals. This implies that the local temperature is below the freezing point of larger crystals, but above the freezing point of the smaller ones, and this reflects the radius dependence of the local surface free energy. In the presence of AF(G)Ps, recrystallization is effectively inhibited, since residual ice growth between adsorbed AF(G)P molecules makes the surface curvature crystal size independent and protein spacing dependent, thereby equalizing the local freezing points on the surface of all crystals.

The unique property of AF(G)Ps is the thermal hysteresis activity (TH), which is defined by the separation of melting point and non-colligative freezing point that creates a temperature window in which ice crystals neither grow nor melt. Protein-ice adsorption and the Kelvin effect are also generally used to explain this activity (Fig. 3). From a completely crystallized solution, the colligative melting or freezing point (Tm) is determined by melting, and on a single crystal level one can envision the crystal fronts recessing with near zero curvature (Fig. 3A). Cooling the solution slightly below Tm, melting ceases and protein adsorption occurs. In the simplest model (Fig. 3C, *iii*) bound AF(G)P molecules confine further ice growth to the area between the AF(G)Ps upon further cooling, which results in an increased surface area and curvature. The surface becomes less stable relative to a planar interface, and the local equilibrium freezing point is lowered. Growth will completely cease once the surface curvature leads to a local freezing point depression equivalent to the supercooling of the solution. Further cooling then, while leading to an increase in local curvature, will not result in observable macroscopic growth until the non-colligative freezing point (To) is reached. This burst point is envisioned to be characterized by a decrease of the critical radius for ice crystal growth falling below the largest AF(G)P spacing, leading to the stochastic inclusion of AF(G)P and initially local unrestricted freezing. This creates a new ice front that cannot be controlled given the large degree of undercooling and results in complete freezing, although altered ice growth kinetics and morphology establish further AF(G)P adsorption to the growing ice front.

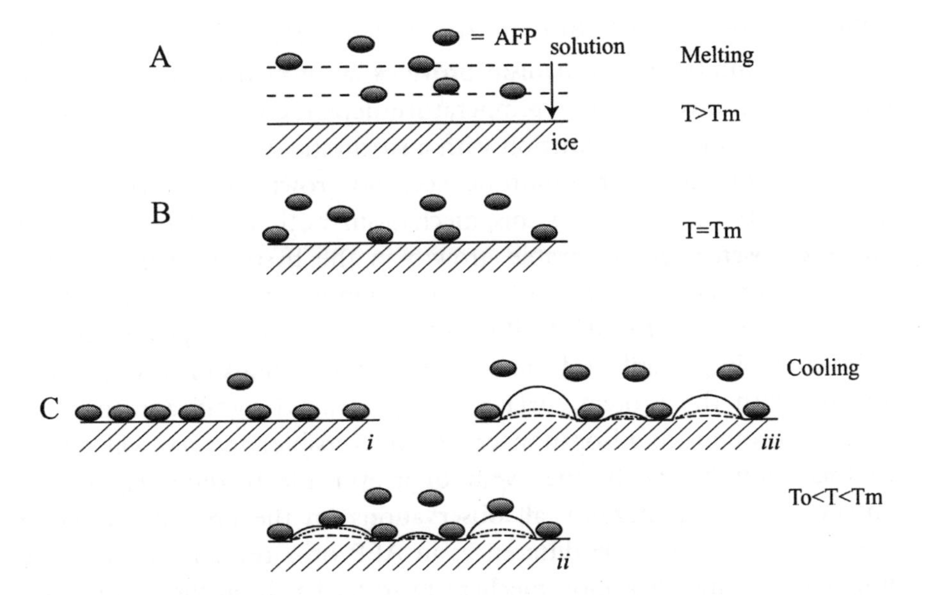

Fig. 3 Model of AFP adsorption and ice inhibition mechanism. **(A)** Above the melting temperature Tm the ice front recedes. **(B)** At this equilibrium transition point, AFP molecules recognize and adsorb to their specific binding sites on the ice crystal surface. **(C)** Irreversible AFP binding competes with crystal growth which depends on degree and rate of undercooling. Fast (i), medium (ii), and slow (iii) protein adsorption rates correlate with expression of further AFP-binding sites, surface coverage and regularity, while the ice surface curvature depends on the degree of undercooling (–––, ---, — indicating decreasing temperatures). The coverage ultimately determines the burst point To, at which the rapid protein incorporation and uncontrollable crystal growth occurs.

Before we critically analyze the limitations of this mechanism, it should be emphasized that the specific molecular events at the ice surface that characterize the three phases of the TH assay are not known: the crystal burst at the non-colligative freezing points, the complete absence of observable crystal growth during the thermal hysteresis window, and the initial, rapid conversion of the crystal into the regular, hexagonal bipyramids. Thus far, any proposed mechanism remains largely speculative, and it is only recently that STM provided molecular level visualization of an ice-surface in the presence of type I AFP, that could serve as direct evidence for the increased ice-curvature between

AFP, and the applicability of the Kelvin effect.[78] Certainly, the adsorption-inhibition mechanism provides an attractive and descriptive explanation, invoking the temperature-dependent critical radius or surface curvature in conjunction with a stochastic AF(G)P incorporation to explain the sudden transition between no growth and uncontrollable growth at To. However, in this mechanism both the burst point To and the absence of observable growth in the hysteresis window lead to the requirement for irreversible protein binding to the crystal surface. The knowledge of the AF(G)P structure, the nature of protein-ligand interactions in general, and the competing interactions of solvent water and ice with the protein surface, though, raised significant concerns whether protein-ice adsorption can be irreversible, and we will return to this point below. In the event of irreversible binding, e.g. in the absence of an equilibrium, all observations on the growth inhibition need to be explained by differences in AFP on-rates, and the primary limitation of the adsorption mechanism so far has been the explanation of the concentration dependence of the thermal hysteresis activity. In this context, it is helpful to focus on the initial phase of this assay. The rapid, seemingly instantaneous regularization of the crystal shape upon initial undercooling indicates that protein adsorption can kinetically compete with crystal growth. Thus, after binding of the initial AF(G)P, further irreversible AF(G)P adsorption will occur. Because AF(G)Ps retard crystal growth most effectively in their adsorption plane (as in this plane the protein has the highest concentration[23]), this will lead to enhanced expression of the adsorption plane and additional AF(G)P-binding (Fig. 3C). The faster the AF(G)P binds, the more ice curvature growth will be inhibited by subsequent AF(G)P-binding, and the higher the surface coverage will be. This concept of ice-shaping has been proposed to be present also at sub-molecular dimensions.[80] Since the protein adsorption rate is concentration dependent, these considerations introduce a concentration dependent coverage. However, in essence the properties of the crystals, i.e. surface regularity, coverage and AF(G)P spacing, would be rapidly set in the initial stage of the cooling. Therefore, this explanation as well as an alternative hypothesis relating the concentration dependence to processes near the burst-point,[11] fail to explain the observed cooling rate dependencies of TH measurements, which suggest the presence of kinetic or equilibrium changes at the AF(G)P-ice surface at all temperatures in the hysteresis gap.

Accepting such changes, however, would invoke the same argument that is forwarded against reversible ice-binding of AF(G)P. Protein desorption or protein incorporation into the ice lattice at temperatures between melting and burst point are perceived to invariably lead to ice growth owing to the much more rapid adsorption of the 55M water relative to the (re-)adsorption of AF(G)P. Thus, growth would not only eventually become macroscopically observable, but also endanger the survival of fish that live in undercooled waters for years. While this appears to be a powerful argument, it is essentially only a lack of mechanistic ideas to describe the surprising growth inhibition (rather than retardation) that has helped the adsorption-inhibition mechanism and irreversible-binding concept to persist against the experimental evidence that is compatible with, or even requires reversible binding. For example, irreversible binding should lead to incorporation of all bound AF(G)P into the ice upon growth, but measured quantities[75] are lower than the expectations from surface coverage estimates. One could also argue that irreversible adsorption would invariably lead to nearly complete ice surface coverage (as shown in Fig. 3C). Further, in many mutant AFPs lowered burst points are also accompanied with residual growth, and less sudden or absent burst-points.[42,81] And foremost, thermal hysteresis activity is concentration dependent and can largely be fit to equilibrium adsorption.[82] All these observations suggest the presence of concentration dependent processes at temperatures within the TH gap that might be slow but still present for even the fully active AFPs. Consequently, alternative mechanistic models for the ice-growth inhibition of AFPs have been proposed using concepts such as a negative line excess,[83] and anisotropic interfacial energies.[84] Nevertheless, it appears that new mechanistic concepts will be measured by their ability to explain the absence of observable crystal growth in the TH gap in order to challenge the popular adsorption-inhibition mechanism successfully.

Mainly in the foreground on the assumption of irreversible protein adsorption, the knowledge of the protein properties and structures has failed to lead to the emergence of a single accepted mechanism explaining the AF(G)P affinity and specificity for ice binding. As outlined, AF(G)P structures display large divergence in secondary and tertiary structure. From a functional view, more surprising are the concurrent dissimilarities of the backbone and surface dynamic properties. Both an increase in backbone stability (as shown by the correlation of helicity and activity for type I AFPs) and rotational restriction of side chain would be expected to increase ice-surface recognition and loss of conformational entropy upon ice-binding. However, the concept of the rigidity of the proteins ice-binding site still applies if we consider the partially lower activities of the apparent exceptions AFGP and type I AFP, respectively. Conformational entropies are still reduced in the molecules by the facts that sugar units provides AFGP with somewhat rigid "side chain," while alanine side chains, which comprise at least 30% if not >60% of the type I AFP-binding site, are not experiencing conformational entropy loss upon adsorption at all. Despite the current limits and ambiguity on the exact nature of the ice-binding site of several AFPs, apparent common surface properties are their planarity and a relatively high nonpolar character. A regular spacing of polar and presumably nonpolar groups, although not always readily apparent, is perceived to not only facilitate a match of polar groups, but a general surface complementarity between the corrugated protein and specific ice surfaces. The latter conclusion, while well-founded by the significant steric sensitivities of AFP ice-binding, is surprising given the nature of the ice surface as being highly dynamic with a transition area of approx. 15 to 20 Å between ice and bulk water.[85–87] Further, no consensus exists in explaining the energetic source of AF(G)P ice-binding. Mechanisms traditionally focused exclusively on H-bonding. The significant activity of type I AFP analogs, in which Thr residues were replaced by nonpolar analogs challenge these views. Further, it has been argued that the number of hydrogen bonds not only are relatively low, but that in comparison with competing protein-water H-bonding, protein-ice H-bonds are weaker due to the steric restrictions of the interacting groups on both sides of the interface.[44,62] This argument, however, focuses on individual hydrogen bonds, and neglects the cooperativity in the formation of multiple bonds being formed simultaneously, which leads to additional entropic contributions or, in other words, increased local concentrations. Thus, it seems possible that despite less ideal protein-ice H-bond geometry and an unfavorable enthalpy contribution, protein-ice H-bonding potentially leads to an overall favorable contribution to the free energy of the AF(G)P adsorption, if most or all protein and ice groups in the interface were to H-bond. Nevertheless, the focus on specific groups is not necessary and in our opinion unfounded, given that the protein surface comprises much more than only H-bonding groups. The surface flatness and sensitivity to steric mutations suggest direct or indirect contributions to the protein-ice affinity of all surface atoms, polar and nonpolar alike and van der Waals interactions between nonpolar protein groups and ice have been proposed.[62,88] Although not generally accepted, they certainly will be present and can in principle be larger than those of protein with water. One could argue that the intimate contact between protein and ice facilitated by the surface complementarity creates in effect a larger density of atoms at the interface. Another, potentially larger contribution may arise from solvation effects[62] that in the context of hydrophobic solutes and proteins is commonly described as hydrophobic interactions. Both polar and nonpolar groups in the protein surface, and ice have a structuring effect on solvent water. The adsorption process, again based on the intimate and extensive surface contact, leads to a desolvation of protein and ice surfaces, and the release of the surface bound water results in an entropy increase. It should be noted that invoking hydrophobic interactions does not imply that ice is less polar than water, but refers to the noted ordering effect of the ice surface that extends nearly 20 Å into the solution. Thus, one can similarly envision a favorable contribution from the ice surface desolvation. It is particularly interesting that the diffusive nature of the ice water interface results in this contribution not being directly proportional to the area of the protein-ice binding site, but a dependence on the volume of the protein that is placed in the ice-water interface. This significantly increases the size of this contribution, thus potentially resulting in the entropic gain of the total displaced water being the dominant factor in the protein-ice adsorption. This hypothesis further presents an explanation for the steric sensitivity of AF(G)P adsorption, as it introduces an energetic driving force for the protein to traverse the ice-water interface and bind at its solid side. Experimental support can be drawn from the antifreeze activity of AFP-antibody complexes[89] and AFP-fusion proteins.[90] The increased volume of these AFPs placed in the interface would lead to increased protein-ice affinity and increased activity at all protein concentrations in this mechanistic framework, without requiring intimate contact between ice and the non-ice binding antibody or non-AFP moieties.

Conclusions

The many structural properties of AF(G)Ps outlined within the context of this review are significant with regards to these protein's function of inhibiting ice formation. Although unrelated, the five types of AF(G)Ps exhibit a few similar properties defining their ice-binding capability evidenced by thermal hysteresis and recrystallization inhibition studies. Many important findings based on biochemical and structural analysis have described some of the essential aspects of protein-ice interactions responsible for the activity of all AF(G)Ps, and challenged previously accepted mechanistic frameworks such as hydrogen-bonding and the prevalent adsorption inhibition mechanism. However, alternative concepts have still to emerge and find general acceptance, and to date the details of AF(G)P function remain enigmatic. The full potential of the structural diversity is yet to be exploited, but should ultimately lead to a better understanding of this fascinating and unique group of proteins, unless their mechanism of action is unexpectedly as diverse as the AF(G)P molecular nature itself.

References

1. DeVries AL, Vandenheede J and Feeney RE (1971). Primary structure of freezing point-depressing glycoproteins. *J. Biol. Chem.* **246**: 305–308.
2. Scholander PF, van Dam L, Kanwisher JW, Hammel HT and Gordon MS (1957). Supercooling and osmoregulation in Arctic fish. *J. Cell. Comp. Physiol.* **49**: 5–24.

3. Duman JG and DeVries AL (1972). Freezing behavior of aqueous solutions of glycoproteins from the blood of an Antarctic fish. *Cryobiology* **9**: 469–472.
4. Davies PL and Hew CL (1990). Biochemistry of fish antifreeze proteins. *FASEB J.* **4**: 2460–2468.
5. Knight CA, Wen D and Laursen RA (1995). Nonequilibrium antifreeze peptides and the recrystallization of ice. *Cryobiology* **32**: 23–34.
6. Carpenter JF and Hansen TN (1992). Antifreeze protein modulates cell survival during cryopreservation: mediation through influence on ice crystal growth. *Proc. Natl. Acad. Sci. USA* **89**: 8953–8957.
7. Chao H, Davies PL and Carpenter JF (1996). Effects of antifreeze proteins on red blood cell survival during cryopreservation. *J. Expt. Biol.* **199**: 2071–2076.
8. Hays LM, Feeney RE, Crowe LM, Crowe JH and Oliver AE (1996). Antifreeze glycoproteins inhibit leakage from liposomes during thermotropic phase transitions. *Proc. Natl. Acad. Sci. USA* **93**: 6835–6840.
9. Lin Q, Ewart KV, Yang DS and Hew CL (1999). Studies of a putative ice-binding motif in winter flounder skin-type antifreeze polypeptide. *FEBS Lett.* **453**: 331–334.
10. Low WK, Miao M, Ewart KV, Yang DS, Fletcher GL and Hew CL (1998). Skin-type antifreeze protein from the shorthorn sculpin, *Myoxocephalus scorpius*: expression and characterization of a Mr 9,700 recombinant protein. *J. Biol. Chem.* **273**: 23098–23103.
11. Fletcher G, Hew C and Davies P (2001). Antifreeze proteins of teleost fishes. *Ann. Rev. Physiol.* **63**: 359–390.
12. Liou YC, Tocilj A, Davies PL and Jia Z (2000). Mimicry of ice structure by surface hydroxyls and water of a beta-helix antifreeze protein. *Nature* **406**: 322–324.
13. Graether SP, Kuiper MJ, Gagne SM, Walker VK, Jia Z, Sykes BD and Davies PL (2000). Beta-helix structure and ice-binding properties of a hyperactive antifreeze protein from an insect. *Nature* **406**: 325–328.
14. Davies PL and Sykes BD (1997). Antifreeze proteins. *Curr. Opin. Struct. Biol.* 7: 828–834.
15. DeVries AL (1971). Glycoproteins as biological antifreeze agents in Antarctic fishes. *Science* **172**: 1152–1155.
16. Yeh Y and Feeney RE (1996). Antifreeze proteins: structures and mechanism of function. *Chem. Rev.* **96**: 601–616.

17. Drewes JA and Rowlen KL (1993). Evidence for a gamma-turn motif in antifreeze glycopeptides. *Biophys. J.* **65**: 985–991.
18. Bush CA and Feeney RE (1986). Conformation of the glycotripeptide repeating unit of antifreeze glycoprotein of polar fish as determined from the fully assigned proton NMR spectrum. *Int. J. Pept. Protein Res.* **28**: 386–397.
19. Filira F, Biondi L, Scolaro B, Foffani MT, Mammi S, Peggion E and Rocchi R (1990). Solid phase synthesis and conformation of sequential glycosylated polytripeptide sequences related to antifreeze glycoproteins. *Int. J. Biol. Macromol.* **12**: 41–49.
20. Rao BN and Bush CA (1987). Comparison by 1H-NMR spectroscopy of the conformation of the 2600 dalton antifreeze glycopeptide of polar cod with that of the high molecular weight antifreeze glycoprotein. *Biopolymers* **26**: 1227–1244.
21. Lane AN, Hays LM, Feeney RE, Crowe LM and Crowe JH (1998). Conformational and dynamic properties of a 14-residue antifreeze glycopeptide from Antarctic cod. *Protein Sci.* **7**: 1555–1163.
22. Lane AN, Hays LM, Tsvetkova N, Feeney RE, Crowe LM and Crowe JH (2000). Comparison of the solution conformation and dynamics of antifreeze glycoproteins from Antarctic fish. *Biophys. J.* **78**: 3195–3207.
23. Knight CA and DeVries AL (1994). Effects of a polymeric, nonequilibrium "antifreeze" upon ice growth from water. *J. Cryst. Growth* **143**: 301–310.
24. Ahmed AI, Yeh Y, Osuga YY and Feeney RE (1976). Antifreeze glycoproteins from Antarctic fish. Inactivation by borate. *J. Biol. Chem.* **251**: 3033–3036.
25. Feeney RE, Osuga DT and Yeh Y (1991). Effect of boronic acids on antifreeze proteins. *J. Protein Chem.* **10**: 167–170.
26. Ben R (2001). Antifreeze glycoproteins: preventing the growth of ice. *Chem. Bio. Chem.* **2**: 161–166.
27. Ben RN, Eniade AA and Hauer L (1999). Synthesis of a C-linked antifreeze glycoprotein (AFGP) micmic: probes for investigating the mechanism of action. *Org. Lett.* **1**: 1759–1762.
28. Tsuda T and Nishimura SI (1996). Synthesis of an antifreeze glycoprotein analogue: efficient preparation of sequential glycopeptide polymers. *Chem. Commun.* 2779–2780.

29. Deng G, Andrews DW and Laursen RA (1997). Amino acid sequence of a new type of antifreeze protein, from the longhorn sculpin *Myoxocephalus octodecimspinosis*. *FEBS Lett.* **402**: 17–20.
30. Zhao Z, Deng G, Lui Q and Laursen RA (1998). Cloning and sequencing of cDNA encoding the LS-12 antifreeze protein in the longhorn sculpin, *Myoxocephalus octodecimspinosis*. *Biochim. Biophys. Acta* **1382**: 177–180.
31. Sicheri F and Yang DSC (1995). Ice-binding structure and mechanism of an antifreeze protein from winter flounder. *Nature* **375**: 427–431.
32. Laursen RA, Wen D and Knight CA (1994). Enantioselective Adsorption of the D- and L-forms of an α-helical antifreeze polypeptide the {2 0 $\bar{2}$ 1} Planes of Ice. *J. Am. Chem. Soc.* **116**: 12057–12058.
33. Knight CA, Cheng CC and DeVries AL (1991). Adsorption of alpha-helical antifreeze peptides on specific ice crystal surface planes. *Biophys. J.* **59**: 409–418.
34. Raymond JA, Wilson PW and DeVries AL (1989). Inhibition of growth of nonbasal planes in ice by fish antifreezes. *Proc. Natl. Acad. Sci. USA* **86**: 881–885.
35. Wen D and Laursen RA (1992). Structure-function relationships in an antifreeze polypeptide: the role of neutral, polar amino acids. *J. Biol. Chem.* **267**: 14102–14108.
36. Wen D and Laursen RA (1992). A model for binding of an antifreeze polypeptide to ice. *Biophys. J.* **63**: 1659–1662.
37. Gronwald W, Chao H, Reddy DV, Davies PL, Sykes BD and Sönnichsen FD (1996). NMR characterization of side chain flexibility and backbone structure in the type I antifreeze protein at near freezing temperatures. *Biochemistry* **35**: 16698–16704.
38. Chao H, Houston ME Jr, Hodges RS, Kay CM, Sykes BD, Loewen MC, Davies PL and Sönnichsen FD (1997). A diminished role for hydrogen bonds in antifreeze protein binding to ice. *Biochemistry* **36**: 14652–14660.
39. Haymet AD, Ward LG, Harding MM and Knight CA (1998). Valine substituted winter flounder "antifreeze": preservation of ice growth hysteresis. *FEBS Lett.* **430**: 301–306.
40. Zhang W and Laursen RA (1998). Structure-function relationships in a type I antifreeze polypeptide. The role of threonine methyl and hydroxyl groups in antifreeze activity. *J. Biol. Chem.* **273**: 34806–34812.

41. Haymet AD, Ward LG and Harding MM (2001). Hydrophobic analogues of the winter flounder "antifreeze" protein. *FEBS Lett.* **491**: 285–288.
42. Loewen MC, Chao H, Houston ME Jr, Baardsnes J, Hodges RS, Kay CM, Sykes BD, Sönnichsen FD and Davies PL (1999). Alternative roles for putative ice-binding residues in type I antifreeze protein. *Biochemistry* **38**: 4743–4749.
43. Baardsnes J, Kondejewski LH, Hodges RS, Chao H, Kay C and Davies PL (1999). New ice-binding face for type I antifreeze protein. *FEBS Lett.* **463**: 87–91.
44. Haymet AD, Ward LG and Harding MM (1999). Winter flounder "antifreeze" proteins: synthesis and ice growth inhibition of analogues that probe the relative importance of hydrophobic and hydrogen-bonding interactions. *J. Am. Chem. Soc.* **121**: 941–948.
45. Zhang W and Laursen RA (1999). Artificial antifreeze polypeptides: alpha-helical peptides with KAAK motifs have antifreeze and ice crystal morphology modifying properties. *FEBS Lett.* **455**: 372–376.
46. Wierzbicki A, Knight CA, Rutland TJ, Muccio DD, Pybus BS and Sikes CS (2000). Structure-function relationship in the antifreeze activity of synthetic alanine-lysine antifreeze polypeptides. *Biomacromolecules* **1**: 268–274.
47. Wierzbicki A, Taylor MS, Knight CA, Madura JD, Harrington JP and Sikes CS (1996). Analysis of shorthorn sculpin antifreeze protein stereospecific binding to $(2\ \bar{1}\ 0)$ faces of ice. *Biophys. J.* **71**: 8–18.
48. Ewart KV, Rubinsky B and Fletcher GL (1992). Structural and functional similarity between fish antifreeze proteins and calcium-dependent lectins. *Biochem. Biophys. Res. Commun.* **185**: 335–340.
49. Ng NF and Hew CL (1992). Structure of an antifreeze polypeptide from the sea raven: disulfide bonds and similarity to lectin-binding proteins. *J. Biol. Chem.* **267**: 16069–16075.
50. Sönnichsen FD, Sykes BD and Davies PL (1995). Comparative modeling of the three-dimensional structure of type II antifreeze protein. *Protein Sci.* **4**: 460–471.
51. Ewart KV, Yang DSC, Ananthanarayanan VS, Fletcher GL and Hew CL (1996). Ca^{2+}-dependent antifreeze proteins: modulation of conformation and activity by divalent metal ions. *J. Biol. Chem.* **271**: 16627–16632.

52. Gronwald W, Loewen MC, Lix B, Daugulis AJ, Sönnichsen FD, Davies PL and Sykes BD (1998). The solution structure of type II antifreeze protein reveals a new member of the lectin family. ***Biochemistry*** **37**: 4712–4721.
53. Ewart KV, Li Z, Yang DSC, Fletcher GL and Hew CL (1998). The ice-binding site of atlantic herring antifreeze protein corresponds to the carbohydrate-binding site of C-type lectins. ***Biochemistry*** **37**: 4080–4085.
54. Cheng CC and DeVries AL (1991). The role of antifreeze glycopeptides and peptides in the freezing avoidance of cold-water fish. In: di Prisco G (ed.), *Life Under Extreme Conditions*. Springer-Verlag, Berlin, Heidelberg, pp. 1–14.
55. Wierzbicki A, Madura JD, Salmon C and Sönnichsen FD (1997). Modeling studies of binding of sea raven type II antifreeze protein to ice. *J. Chem. Inf. Comput. Sci.* **37**: 1006–1010.
56. Loewen MC, Gronwald W, Sönnichsen FD, Sykes BD and Davies PL (1998). The ice-binding site of sea raven antifreeze protein is distinct from the carbohydrate-binding site of the homologous C-type lectin. *Biochemistry* **37**: 17745–11753.
57. Bertrand JA, Pignol D, Bernard JP, Verdier JM, Dagorn JC and Fontecilla-Camps JC (1996). Crystal structure of human lithostathine, the pancreatic inhibitor of stone formation. *EMBO J.* **15**: 2678–2684.
58. Hew CL, Slaughter D, Joshi SB and Fletcher GL (1984). Antifreeze polypeptides from the Newfoundland ocean pout, *Macrozoarces americanus*: presence of multiple and compositionally diverse components. *J. Comp. Physiol. B* **155**: 81–88.
59. Hew CL, Wang NC, Joshi S, Fletcher GL, Scott GK, Hayes PH, Buettner B and Davies PL (1988). Multiple genes provide the basis for antifreeze protein diversity and dosage in the ocean pout, *Macrozoarces americanus*. *J. Biol. Chem.* **263**: 12049–12055.
60. Schrag JD, Cheng CH, Panico M, Morris HR and DeVries AL (1987). Primary and secondary structure of antifreeze peptides from Arctic and Antarctic zoarcid fishes. *Biochim. Biophys. Acta* **915**: 357–370.
61. Sönnichsen FD, Sykes BD, Chao H and Davies PL (1993). The non-helical structure of antifreeze protein type III. *Science* **259**: 1154–1157.

62. Sönnichsen FD, DeLuca CI, Davies PL and Sykes BD (1996). Refined solution structure of type III antifreeze protein: hydrophobic groups may be involved in the energetics of the protein-ice interaction. *Structure* **4**: 1325–1337.
63. Jia Z, DeLuca CI, Chao H and Davies PL (1996). Structural basis for the binding of a globular antifreeze protein to ice. *Nature* **384**: 285–288.
64. Miura K, Ohgiya S, Hoshino T, Nemoto N, Suetake T, Miura A, Spyracopoulos L, Kondo H and Tsuda S (2001). NMR analysis of type III antifreeze protein intramolecular dimer: structural basis for enhanced activity. *J. Biol. Chem.* **276**: 1304–1310.
65. Antson AA, Smith DJ, Roper DI, Lewis S, Caves LS, Verma CS, Buckley SL, Lillford PJ and Hubbard RE (2001). Understanding the mechanism of ice binding by type III antifreeze proteins. *J. Mol. Biol.* **305**: 875–889.
66. Yang DSC, Hon WC, Bubanko S, Xue Y, Seetharaman J, Hew CL and Sicheri F (1998). Identification of the ice-binding surface on a type III antifreeze protein with a "flatness function" algorithm. *Biophys. J.* **74**: 2142–2151.
67. Sönnichsen FD, Davies PL and Sykes BD (1998). NMR structural studies on antifreeze proteins. *Biochem. Cell Biol.* **76**: 284–293.
68. Chao H, Sönnichsen FD, DeLuca CI, Sykes BD and Davies PL (1994). Structure-function relationship in the globular type III antifreeze protein: identification of a cluster of surface residues required for binding to ice. *Protein Sci.* **3**: 1760–1769.
69. Graether SP, DeLuca CI, Baardsnes J, Hill GA, Davies PL and Jia Z (1999). Quantitative and qualitative analysis of type III antifreeze protein structure and function. *J. Biol. Chem.* **274**: 11842–11847.
70. Chao H, DeLuca CI and Davies PL (1995). Mixing antifreeze protein types changes ice crystal morphology without affecting antifreeze activity. *FEBS Lett.* **357**: 183–186.
71. DeLuca CI, Davies PL, Ye Q and Jia Z (1998). The effects of steric mutations on the structure of type III antifreeze protein and its interaction with ice. *J. Mol. Biol.* **275**: 515–525.
72. Westh P, Ramlov H, Wilson PW and DeVries AL (1997). Vapour pressure of aqueous antifreeze glycopeptide solutions. *Cryo. Lett.* **18**: 277–282.

73. Haschemeyer AE, Cuschbauer W and DeVries AL (1977). Water binding by antifreeze glycoprotins from Antarctic fish. *Nature* **269**: 87–88.
74. Duman JG, Patterson JL, Kozak JJ and DeVries AL (1980). Isopiestic determination of water binding by fish antifreeze glycoproteins. *Biochim. Biophys. Acta* **626**: 332–336.
75. Raymond JA and DeVries AL (1977). Adsorption inhibition as a mechanism of freezing resistance in polar fishes. *Proc. Natl. Acad. Sci. USA* **74**: 2589–2593.
76. Kerr WL, Feeney RE, Osuga DT and Reid DS (1986). Interfacial energies between ice and solutions of anttifreeze glycoproteins. *Cryo. Lett.* **6**: 371–378.
77. Wilson PW, Beaglehole D and DeVries AL (1993). Antifreeze glycopeptide adsorption on single crystal ice surfaces using ellipsometry. *Biophys. J.* **64**: 1878–1884.
78. Grandum S, Yabe A, Nakagomi K, Tanaka M, Fumio T and Kobayashi Y (1999). Analysis of ice crystal growth for a crystal surface containing adsorbed antifreeze proteins. *J. Crystal Growth* **205**: 382–390.
79. Wilson PW (1993). Explaining thermal hysteresis by the Kelvin effect. *Cryo. Lett.* **14**: 31–36.
80. Houston ME Jr, Chao H, Hodges RS, Sykes BD, Kay CM, Sönnichsen FD, Loewen MC and Davies PL (1998). Binding of an oligopeptide to a specific plane of ice. *J. Biol. Chem.* **273**: 11714–11718.
81. DeLuca CI, Chao H, Sönnichsen FD, Sykes BD and Davies PL (1996). Effect of type III antifreeze protein dilution and mutation on the growth inhibition of ice. *Biophys. J.* **71**: 2346–2355.
82. Burcham TS, Osuga DT, Yeh Y and Feeney RE (1986). A kinetic description of antifreeze glycoprotein activity. *J. Biol. Chem.* **261**: 6390–6397.
83. Hall DG and Lips A (1999). Phenomenology and mechanism of antifreeze peptide activity. *Langmuir* **15**: 1905–1912.
84. Wilson PW (1994). A model for thermal hysteresis utilizing the anisotropic interfacial energy of ice crystals. *Cryobiology* **31**: 406–412.
85. Karim OA, Kay PA and Haymet ADJ (1988). The ice/water interface: a molecular dynamics simulation using the simple point charge model. *J. Chem. Phys.* **92**: 4634–4635.
86. Hayward JA and Haymet ADJ (2001). The ice/water interface: molecular dynamics simulations of the basal, prism, $\{2\ 0\ \bar{2}\ 1\}$, and $\{2\ \bar{1}\ \bar{1}\ 0\}$ interfaces of ice I_h. *J. Chem. Phys.* **114**: 3713–3726.

87. Dalal P, Knickelbein J, Haymet ADJ, Sönnichsen FD and Madura JD (2001). Hydrogen bond analysis of Type 1 antifreeze protein in water and the ice-/water interface. *Phys. Chem. Comm.* 7: 1–5.
88. Wen D and Laursen RA (1993). Structure-function relationships in an antifreeze polypeptide: the effect of added bulky groups on activity. *J. Biol. Chem.* **268**: 16401–16405.
89. Wu DW, Duman JG and Xu L (1991). Enhancement of insect antifreeze protein activity by antibodies. *Biochim. Biophys. Acta* **1076**: 416–20.
90. DeLuca CI, Comley R and Davies PL (1998). Antifreeze proteins bind independently to ice. *Biophys. J.* **74**: 1502–1508.

Chapter 6

Control of Antifreeze Protein Gene Expression in Winter Flounder

Ming Miao
Department of Biochemistry
University of Toronto
Toronto, Ontario, M5G 1L5, Canada

Shing-Leng Chan
Institute of Molecular and Cell Biology
National University of Singapore
30 Medical Drive, Singapore 117609

Garth L Fletcher
Ocean Sciences Center
Memorial University of Newfoundland
St. John's, Newfoundland, A1C 5S7, Canada

Choy L Hew*
Department of Biological Sciences and
Tropical Marine Science Institute
National University of Singapore, Singapore

Introduction

Winter flounder (*Pleuronectus americanus*) resides along the Atlantic coast from Labrador, Canada, to Georgia, USA. This species frequently encounters seawater temperatures below its own freezing point (–0.8°C) during winter months in the northern ranges. Therefore, it produces antifreeze proteins (AFPs). The antifreeze proteins of the Newfoundland winter flounder populations (wfAFPs) are under tight seasonal and hormonal regulation, mainly at the transcriptional level. Furthermore, two sets of wfAFPs isoforms encoded by distinct genes with different tissue distributions and differential expression have been reported. These proteins and their corresponding gene families are an excellent model for studying the mechanism of transcriptional regulation mediated by environmental and endocrine factors. This article focuses on the control of wfAFP gene expression with respect to its seasonal and hormonal regulation. The mechanisms for its tissue specificity, as well as the specific transcription factors and cis-elements will be discussed.

Biochemistry of the wfAFPs

The wfAFPs belong to the type I AFPs, which are alanine-rich and α-helical polypeptides (for review: Harding *et al.* (1990)[1] and Chapter 7 in this monograph). Winter flounder is the first species found to produce two distinct isoforms of type I AFPs: the liver-type AFPs (wflAFPs) and the skin-type AFPs (wfsAFPs).

The Liver-Type wfAFPs

The wflAFPs are synthesized mainly in the liver as large precursor polypeptides of 82 amino acids. The presequences are removed cotranslationally while the prosequences are removed soon after their secretion into blood.[2–4] The proteins were resolved by reverse phase high performance liquid chromatography (HPLC) into several components and two of them (HPLC-6 and 8, or wflAFP-6 and 8) make up to 90% of the plasma wflAFPs.[5,6] These major wflAFPs are 37 amino acids in length and contain three 11 amino acid repeats of Thr-X_2-Asp/Asn-X_7 where X is predominantly Ala, and the tertiary structure of wflAFP-6 has been elucidated as a single α-helix.[7,8] Flounder AFPs are found to be encoded by a multiple family of genes. Some of them are arrayed in direct tandem repeats and the rest are linked but irregularly spaced.[9] Approximately 40 copies of these AFP genes encode the wflAFPs and the majority of genes from the tandem repeats sequenced so far code for the wflAFP-6 component.[10]

The gene organization of a representative wflAFP gene, 2A-7b, which encodes the wflAFP-6 component, is shown in Fig. 1. It is a tandemly repeated gene less than 1 kb in length and consisting of two exons and a single intron of 496 bp.[10,11] It contains a mRNA cap site, TATA and CAAT boxes which are 32 bp and 84 bp upstream from the cap site, respectively, as well as initiation and stop codons, a conventional splice junction sequences and a polyadenylation signal.

The Skin-Type wfAFPs

Recently, wfAFP mRNAs have been found in many tissues in addition to liver, including skin, scale, fin and gill.[12] Analysis of the wfAFPs isolated from flounder skin indicates that they are also enriched with alanine and contain similar threonine 11-amino acid repeats, while with less antifreeze activity compared to the serum AFPs.[13] wfsAFPs are mostly polypeptides of 37 to 40 amino acids with a blocking acetyl group at the N-terminus. Those proteins are produced as mature polypeptides without the pre- and prosequences, indicating that they may be intracellular proteins. There are also approximately 40 copies of wfsAFP genes in flounder genome which are distinct from the genes encoding wflAFPs.[13] The sequences of wfsAFP cDNA were found to be closely related to two previously identified genomic sequences F2 and 11–3, which belong to the subset of AFP genes that are linked but irregularly spaced. F2 and 11–3 were initially assigned as pseudogenes since they lack a typical TATA box in the putative promoter region and contain stop codons in the 5' upstream region.[14] Two transcriptional start sites were mapped on the F2 gene and a putative TFIID binding motif AATAAAT is found 25 bp upstream of the first start site. Its gene organization is very similar to the liver 2A-7b gene, which also consists of two exons and an intron (Fig. 1). The intron of F2 shares 95% identity to that of 2A-7b except that F2 intron has an extra 241 bp insertion at position +254. The presence of two sets of AFP genes exhibiting distinct tissue specificity within winter flounder has raised interesting questions on how the differential expression is achieved and what DNA element(s) might be involved in this regulation.

Liver-type AFP (2A-7b)

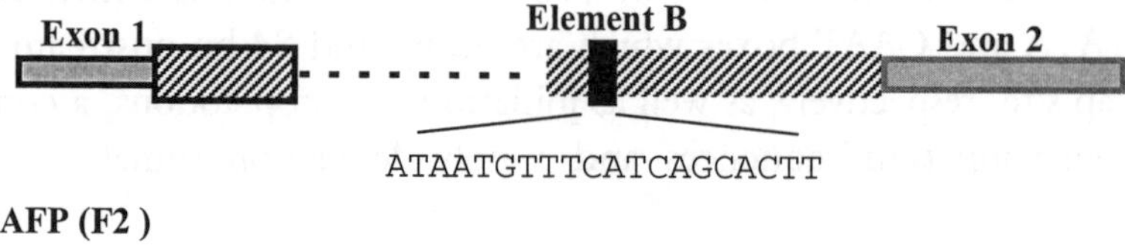

Skin-type AFP (F2)

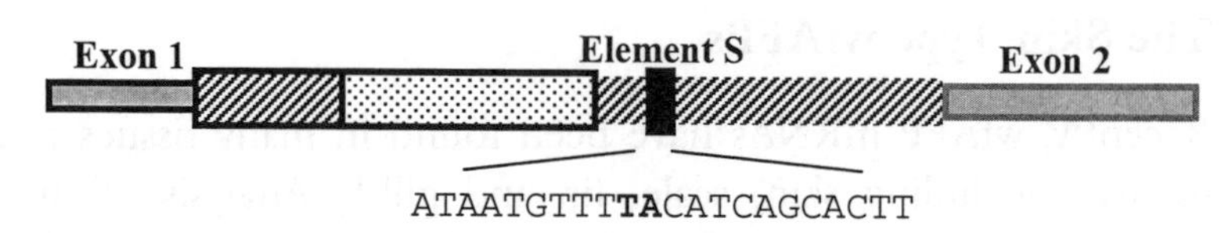

Fig. 1 Organization of the genes encoding wflAFP and wfsAFP. The homologous introns in wflAFP and wfsAFP genes and the additional intron in wfsAFP genes are indicated by striped and dotted boxes, respectively. The figure is not to scale.

Environmental and Hormonal Control of wfAFPs Production

The production of AFPs in winter flounder from the Newfoundland population exhibits dramatic seasonal variations. The serum wfAFPs level, which is reflected by the thermal hysteresis of the fish blood, starts to increase in November and reaches its peak of approximately 5–10 mg/ml in February. Then it starts to decline in May to its minimal level through the summer (Fig. 2).[15,16] Also, the seasonal fluctuations of wfAFP mRNA level in liver match closely, but slightly precede the rise and fall of the AFP level.[17–19] The variations of these AFP mRNAs and protein levels were found to be inversely correlated to water temperature as well as the growth rate of the fish. Numerous studies have been conducted to investigate the environmental and hormonal factors that are implicated in this seasonal production of wfAFPs.

Water Temperature and Photoperiod

Although the variation of the serum wfAFPs correlates with the annual cycle of water temperatures, low temperature does not serve as the

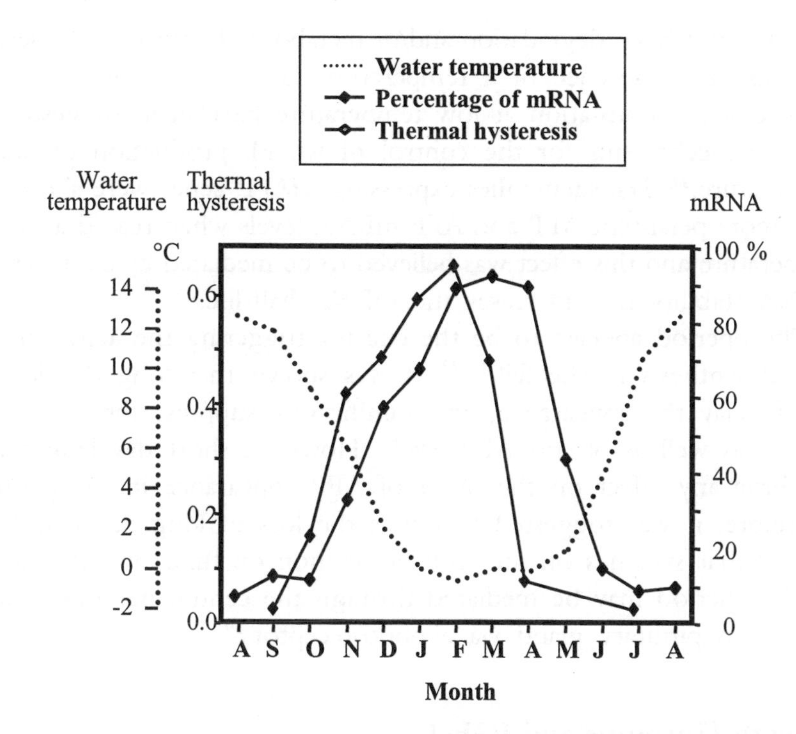

Fig. 2 Seasonal variations of the plasma hysteresis and liver AFP mRNA level in winter flounder. The thermal hysteresis in plasma and the AFP mRNA in liver of the winter flounder are shown with a one-year seasonal cycle. The mRNA levels were calculated as a percentage of peak winter values in February. Adapted with permission from Chan *et al.* (1993) and Fletcher *et al.* (1989).[16,19]

signal to initiate the production of wfAFPs in the fall.[20,21] The AFP mRNA still increased above basal level in the fall when flounders were maintained at 18°C, and fish exposed in 0°C only exhibited a delay of plasma AFP accumulation. Also, the transcription rate of wfAFPs was not affected by water temperature as shown by nuclear run-on experiments.[22] On the other hand, flounders exposed in high water temperatures had low levels of plasma AFPs as well as mRNAs, suggesting that cold temperature is still required for high wfAFP accumulation in the fall. Since the rate of transcription does not vary with the temperature, its effect might be post-transcriptional to alter the rates of mRNA degradation and/or metabolic clearance of the serum proteins. The presence of a temperature-dependent nuclease and/or more efficient translation at low temperature have been suggested as possible mechanisms for the control of wfAFP production by water temperature.[16] Transgenic flies expressing wfAFP genes were shown to have more persistent AFP and AFP mRNA levels when reared at lower temperature and this effect was believed to be mediated by cold-specific mRNA stability that increases the mRNA half-life.[23]

Photoperiod appears to be the cue for triggering the activation of wfAFP synthesis in the fall.[18,21] It was shown that long day length could delay the appearance, and significantly suppress the liver AFP mRNA as well as plasma AFP levels. However, short day length did not have any effect on the onset of AFP appearance in the plasma. Therefore, it was suggested that it is the loss of long day length in the fall that switches on the AFP production on time, and this effect of photoperiod may be mediated through the central nervous system (CNS) on pituitary gland via a photoreceptor.[19]

Growth Hormone and IGF-1

Several studies have demonstrated that the pituitary gland and, specifically, growth hormone (GH) play important roles in controlling the AFP synthesis in winter flounder.[19,24–28] In hypophysectomized flounder, plasma antifreeze activity as well as AFP biosynthesis in the liver remained high during summer. Injection of pituitary extracts significantly reduced the plasma AFP levels in these fish and injection of the GH fraction from the pituitary had similar results. The involvement of GH was further supported by nuclear run-on experiments showing that AFP gene transcription in liver was suppressed in flounder treated with GH. In addition, the seasonal cycle of growth in flounder is inversely correlated with the liver AFP mRNA level. GH mRNA in pituitary also showed a seasonal cycle with the lowest levels occurring during the winter when liver AFP mRNA levels were at their peak (Gill and Fletcher, unpublished data). These results indicate that GH in the pituitary gland plays a direct role in hormonal regulation on the synthesis of serum AFPs in winter flounder.

Insulin-like growth factor (IGF-1) has been known to comprise a major route of GH action. GH stimulates IGF-1 synthesis, mainly from the liver, and consequently increases the circulating IGF-1 level [for review: Daughaday and Rotwein (1989), and Pankor (1999)].[29,30] In teleosts, studies have also shown the liver to be a major site of IGF-1 production in response to GH.[31–33] IGF-1 binds to its receptor (IGF-1R). This leads to phosphorylation of IGF-1R on tyrosine residues and subsequent phosphorylation on insulin receptor substrate-1 (IRS-1) and insulin receptor substrate-2 (IRS-2). Tyrosine phosphorylation of IRS-1 provides binding sites for several distinct Src homology 2 (SH2) proteins and mediates multiple signal pathways including the phosphatidylinositol 3-kinase (PI3-kinase) and Ras/Raf/MAP kinase cascade.[34,35] The involvement of IGF-1 in wfAFP gene expression has been investigated recently. Treatment with IGF-1 has been shown to significantly reduce the core enhancer activity of a liver-type AFP gene in transient expression assays (Miao *et al.*, unpublished data). Moreover, PI3-kinase appears to be one of the downstream factors for IGF-1 signaling since treatment of a PI3-kinase inhibitor, wortmannin, restored the inhibition effect of IGF-1 in transient expression assays. Together, these results suggest that IGF-1 and PI3-kinase have important roles in mediating the effect of GH on wflAFP gene expression.

Differential Regulation of the Skin-Type wfAFPs

The environmental and hormonal studies discussed above were mainly focused on the plasma AFP level and liver AFP mRNAs, i.e. the liver-type AFPs. Therefore, it was interesting to find that the wflAFPs and wfsAFPs are differentially regulated.[36] In contrast to the wflAFP mRNA, which exhibits 500- to 700-fold differences between summer and winter, the mRNA of wfsAFPs only presents a modest six- to ten-fold seasonal variation. Also, while the mRNA of wflAFP in hypophysectomized fish increased over 40-fold, no significant change was found for the wfsAFP upon hypophysectomy, suggesting that the wfsAFP gene is not under GH control. Table 1 summarizes the distinct characteristics between wflAFPs and wfsAFPs.

Table 1 Comparison of the distinct characteristics of wflAFP and wfsAFP.

	Liver-type (wflAFP)	Skin-type (wfsAFP)
Tissue distribution	Liver-specific	Widely expressed, mainly in exterior tissues
Gene organization	Approx. 40 copies, tandem repeats	Approx. 40 copies, linked but irregularly spaced
Seasonal variation	Dramatic, 500- to 700-fold	Moderate, six- to ten-fold
Hormonal control	Inhibited by GH	No significant effect
Biosynthesis	PreproAFP	Mature AFP
Cellular localization	Extracellular	Intracellular

Transcriptional Regulation of the wfAFP Genes

It has been established that transcription is the key mechanism in regulating the seasonal production of wflAFPs.[17,22,27] The AFP mRNA constitutes 0.5% of the total liver RNA during winter and declines to 0.0007% during summer, leading to the rise and fall of plasma AFP levels. Also, the transcriptional rate and accumulation of wflAFP mRNAs in hypophysectomized fish were dramatically reduced by the GH treatment, indicating that the negative GH control is mainly at the transcriptional level. Post-transcriptional regulation, however, may also play a role. One study revealed that the capacity of tRNA acceptors for alanine in winter fish increases 40% over summer fish.[37] Also, the alanyl-tRNA synthetase functions more efficiently at low temperature, indicating that translation efficiency may be controlled by seasonal water temperature.

The wflAFP Intron as a Liver-Specific Enhancer

Transcription of eukaryotic genes is regulated by cis-acting DNA sequences, which may be promoters, enhancers or silencers, and their interactions with DNA-binding proteins. To investigate the regulatory elements of wflAFP genes, the 5' upstream region and the only intron of the 2A-7b gene have been examined for their ability to transactivate the chloramphenicol acetyltransferase (CAT) reporter gene in transient expression.[38,39] Various lengths of the 5'-upstream sequence from 66 bp to –2.3 kb were found to drive similar and very low levels of expression in rainbow trout hepatoma (RTH) cells. On the other hand, the intron of the wflAFP gene (+106 to +602), when linked to the wflAFP basic promoter (–143 to +32), showed more than ten-fold transactivation activity compared to that of the basal promoter in human hepatoma (HepG2) cells. The enhancer activity of the intron has also been demonstrated in flounder hepatocytes by *in vitro* transcription assays.[40] Moreover, it was shown that the intron acts as a liver-specific enhancer since it only functioned in cells of hepatic origin (HepG2), but not non-hepatic cells (HeLa and Rat2). Deletion of the intron sequence from +192 to +334 completely destroyed the enhancer activity and a truncated intron sequence from +192 to +350 exhibited similar activity to the full-length intron. These results indicate that the intron of the 2A-7b wflAFP gene acts as a liver-specific enhancer, and the enhancer region is refined to the sequence +192 to +334 of the gene. In addition, the tissue-specificity of the wflAFP gene was also supported by a study on transgenic salmon (*Salmo salar*) showing that the 2A-7b gene integrated into their genome exhibited liver-specific and seasonal wflAFP expression.[41]

DNA Element and Transcription Factors Involved in the wflAFP Gene Regulation

Regions in the enhancer of wflAFP gene intron that interact with nuclear proteins were mapped by Dnase I footprinting assays with similar results using both rat or flounder liver nuclear extracts.[39,40] One of the regions, designated as Element B (+303 to +322) was shown to form two specific complexes with both rat and flounder extracts by gel retardation assays. One of the complexes could be competed out specifically by oligonucleotide containing C/EBP (CCAAT/enhancer binding protein) binding site. Members of the C/EBP family including C/EBPα, β, γ, δ and ε exhibit similar DNA binding specificity and among them, C/EBPα is expressed only in a limited number of tissues and is present in high concentration in terminally differentiated hepatocytes.[42,43] An antibody specific to C/EBPα was shown to supershift the specific C/EBP complex formed between Element B and rat liver extract. In addition, recombinant rat C/EBPα also formed a specific complex with Element B, while co-transfection of C/EBPα with the wflAFP intron further increased its enhancer activity (Miao *et al.* (1998) and Miao and Hew unpublished data). These studies indicate that the liver-enriched factor C/EBPα interacts with Element B of the wflAFP gene and is involved in its tissue-specific expression. The C/EBP family of proteins has been found from mammals to *Drosophila*, and its presence in flounder has also been demonstrated using antibody against C/EBPβ (cross-reacts to C/EBPα, β, γ and ε).[43–45] As discussed above, the GH regulation of the wflAFP genes is believed to be mediated by IGF-1. IGF-1 was reported to cause C/EBPα dephosphorylation, which is correlated with the repression of GLUT4 transcription in adipocytes.[46] We have proposed that IGF-1, regulated by GH, may similarly cause the deactivation of C/EBPα, hence preventing its interaction to Element B of the enhancer and thus repressing the liver-specific wflAFP gene expression. On the other hand, an oligonucleotide containing an AP1 consensus was found to be able to compete with another specific complex formed between Element B and rat or flounder liver extracts in gel retardation assays. However, the protein that binds to this "AP1 binding site" in Element B was shown to be distinct from the AP1 components (the Jun and Fos family transcription factors[47]). The presence of a novel "AP-1 like" protein was indicated and we proposed it as a presumptive "Antifreeze Enhancer-binding Protein" (AEP) which binds specifically to Element B.

Biochemical studies have been conducted to identify the nucleotides in Element B that are involved in its interactions with nuclear proteins by methylation interference assays. The G residues at position +312 and +320 of the lower strand and the G residues at +308 of the upper strand and +312 of the lower strand of Element B were shown to be important to its binding to C/EBPα and AEP, respectively (Fig. 3).

	Δ : AP1 ∗: C/EBP	Transactivation Activity (%)
Element B WT	+308 Δ ATAATGTTTCATCAGCACTT TATTACAAAGTAGTCGTGAA Δ ∗ +312 ∗ +320	100
m A	ATAATTTTTCATCAGCACTT	6.5
m C	ATAATGTTTCATCAGCAATT	4.3
m A+C	ATAATGTTTAATCAGCACTT	47.8
m 3	ATAATTTTTTATCAGCAATT	3.3
Element S	ATAATGTTTTACATCAGCACTT	28.3

Fig. 3 Interruption of Element B and nuclear protein interactions significantly affects its transactivation ability. Nucleotides important for the interactions between Element B and C/EBPα, as well as the "AP-1 like" protein were mapped by methylation interference assays. Mutations in Element B were introduced to the wflAFP intron (+192–+350) linked to wflAFP gene basic promoter and CAT reporter gene. Transactivation activities were obtained by transient expression assays in HepG2 cells.

Single mutations at +308 or +320 of the truncated liver intron (+192 to +350) of the wfAFP gene were able to destroy its enhancer activity in transient expression assays, while the transactivation activity was significantly reduced when the G nucleotide at +312 was mutated. Together, these findings confirm that Element B is the core enhancer that mediates the transactivation effect of the wflAFP gene intron and the interactions between Element B and nuclear proteins C/EBPα and the presumptive AEP are important for its enhancer activity.

Differential Regulation of the wfsAFP Gene

As discussed above, the skin-type AFP gene is more widely expressed than the liver-specific wflAFP gene, despite the fact that the intron of the wfsAFP gene is homologous to that of the wflAFP gene. Therefore, the function of the wfsAFP intron in transactivation has been investigated.[45] The skin-type intron was found to act as a ubiquitous enhancer. When linked to the CAT reporter gene, the full length wfsAFP intron exhibited transactivation activity in cells of skin origin (newborn rat keratinocytes, NBRK) as well as non-skin origin cells (cervical carcinoma, HeLa). In contrast, the liver-type intron had no enhancer activity in these cell lines.

There is a TA dinucleotide insertion within the region of the skin-type intron, which corresponds to Element B of the liver-type intron (designated as Element S). Element B is known to be important in mediating the enhancer activity of the liver-type intron, thus the effect of the TA insertion in Element S of the wfsAFP gene was studied. Experiments revealed that Element S does not interact with C/EBPα from rat liver extract as strongly as recombinant C/EBPα in gel retardation assays. On the other hand, it still binds specifically to the presumptive AEP. Also, the TA dinucleotide insertion in Element B of the truncated liver-type intron (+192 to +350) dramatically decreases the liver-specific enhancer activity of the wflAFP intron, as demonstrated by transient expression assays in HepG2 cells (Fig. 3). However, this construct exhibited no significant degree of enhancer activity in HeLa and NBRK cells, suggesting that the sequences responsible for skin-type intron activity may reside outside the +192 to +350 region. Together, these findings established the C/EBPα binding motif in Element B mediates the tissue specificity of the wflAFP intron, while the loss of the interaction to C/EBPα in Element S might explain the broader tissue expression pattern of the wfsAFP mRNAs. Also, as discussed above, it was suggested that IGF-1 and C/EBPα mediate the GH regulation of wflAFP genes. The fact that the wfsAFP genes are not under GH regulation may be due to the lack of C/EBPα binding motif in the wfsAFP gene enhancer. A similar disruption of C/EBP binding site has also been reported in the clotting factor IX promoter of haemophilia B patients where it is found to significantly reduce transactivation activity.[48] The participation of C/EBPα and its disruption by naturally occurring mutation may be a common mechanism in controlling differential gene expression in liver. Conversely, the additional 241 bp fragment in the skin-type intron may also be involved in its transcriptional regulation. It is possible that other cis-acting sequences in this skin-type intron fragment interact with transcription factor(s) to dictate the more ubiquitous expression of the wfsAFP gene. Sequences similar to GRE and Oct-1 consensus binding sites were found in these additional fragments of the wfsAFP intron. The role of the putative cis-acting sequences and interacting factors will require further investigation.

Characterization of the AEP

The presence of the presumptive AEP and its interaction with the core enhancer Element B has been demonstrated in both mammalian and flounder hepatocytes. The identity of the AEP analog in rat has been revealed recently by screening a rat liver cDNA expression library via its affinity to Element B as well as RT-PCR.[49] A complete AEP open reading frame sequence was obtained, encoding a novel helicase domain-containing protein with a high percentage of sequence identity to hamster Rip-1, human and mouse Smubp-2 and a truncated human Smubp-2, GF-1.[50–53] The deduced polypeptide sequence of rat AEP is 988 amino acid residues in length with 91.5%, 89.1% and 76.1% identity to its counterparts in mouse, hamster and human, respectively. It contains the seven putative helicase motifs which are characteristic of this family of proteins (Fig. 4). The N-terminal two-thirds of the protein, which includes helicase motifs I–IV, is highly conserved among species. The DNA binding domain contains helicase motifs V and VI and is less conserved.[51] A nuclear localization signal (NLS) is located in the C-terminus portion of the protein. To verify the specific binding of AEB to Element B, a maltose-binding protein and AEP (MBP-AEP) fusion protein was expressed in bacteria. This AEP fusion protein exhibited strong and specific interaction to Element B of the enhancer by Southwestern analysis as well as gel retardation assays. Methylation interference assays showed that the G residues at nucleotide position +308 of the upper strand and +312 of the lower strand of Element

A

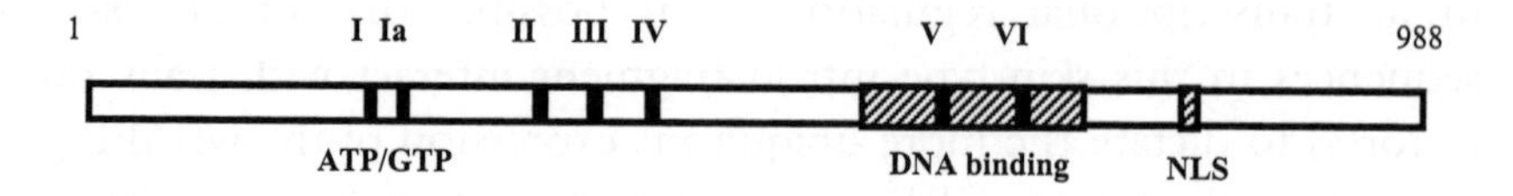

B

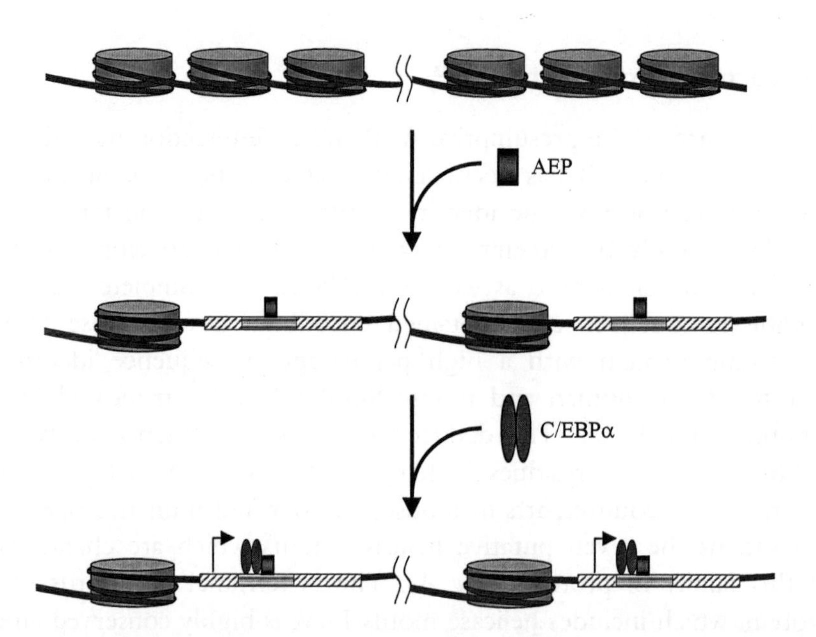

Fig. 4 The rat antifreeze enhancer-binding protein and its possible function in wfAFP gene expression. **(A)** Schematic representation of the rat AEP. The DNA binding domain and the nuclear localization signal (NLS) are shown as striped boxes. The seven putative helicase motifs including the ATP/GTP binding sequences are shown as black boxes. The GenBank accession number for the rat AEP is AF199411. **(B)** The potential role of the AEP in transcription of the tandem repeat wfAFP genes. The genes are packed in chromosome structure when they are inactive. The AEP interacts with the tandemly repeated genes and, through its belicase activity, unwinds the DNA. This may favour the access of other factors such as C/EBPα to DNA binding sites resulting in effective transactivation.

B were important in the interaction with AEP (Miao *et al.*, manuscript submitted). This was consistent with previous methylation interference data for the complex formed between Element B and the "AP1-like" factor in rat liver extract. These results confirmed the identity of rat Antifreeze Enhancer-binding Protein (AEP) and its specific interaction with the enhancer Element B of the wflAFP gene. The AEP family of proteins appears to include ubiquitous factors expressed in all the tissues examined and with genes widely present among species including winter flounder. Moreover, the interaction between AEP and Element B was shown to be required for the activation of the wflAFP gene enhancer in HepG2 cells. A single mutation in Element B at position +308 of the wflAFP gene enhancer was able to interrupt the specific interaction between Element B and AEP and destroy the transactivation activity of the wflAFP enhancer.

The structure of AEP indicates its potential helicase and ATPase activity. DNA helicase can utilize nucleotide triphosphate hydrolysis as energy to break hydrogen bonds between base pairs and unwind the duplex DNA to assist in the action of other enzymes. Therefore, it has been proposed that a protein with DNA helicase activity may enhance transcription by altering chromatin structure, thus facilitating contact between activators and the transcription apparatus.[50,54,55] The flounder AFP genes are arranged in tandem repeats or with multiple linked copies.[9] The chromatin structure of these AFP genes may be altered by the helicase activity of the AEP to promote the access for transcription factors to DNA and achieve rapid high-level gene expression (Fig. 4B). Although the AEP belongs to a family of widely expressed factors, the family includes proteins that are involved in the transcription of many tissue-specific genes including the β-cell-specific insulin II promoter, myocyte-specific atrial natriuretic factor enhancer, glial-specific JCV early and late promoter, and the liver-specific apoA-I promoter.[50,56–58] Studies on the wflAFP genes provided the first evidence that AEP and other tissue-specific factors share an overlapping site of interaction in the enhancer sequence, while the AEP orthologue Smubp-2 was shown to associate with the basal transcription factor TATA binding protein (TBP) to regulate gene expression.[59] The AEP family of proteins may serve as a co-factor that cross-talks between tissue-specific factors and the basic transcription apparatus in the complex mechanism of gene regulation.

Conclusions

A working model has been proposed with a repression mechanism to explain the complex seasonal, hormonal and tissue-specific regulation of the wflAFP genes (Fig. 5). During summer, GH released from the pituitary stimulates the production of IGF-1, which dephosphorylates and deactivates C/EBPα and/or affects the level of AEP, resulting in

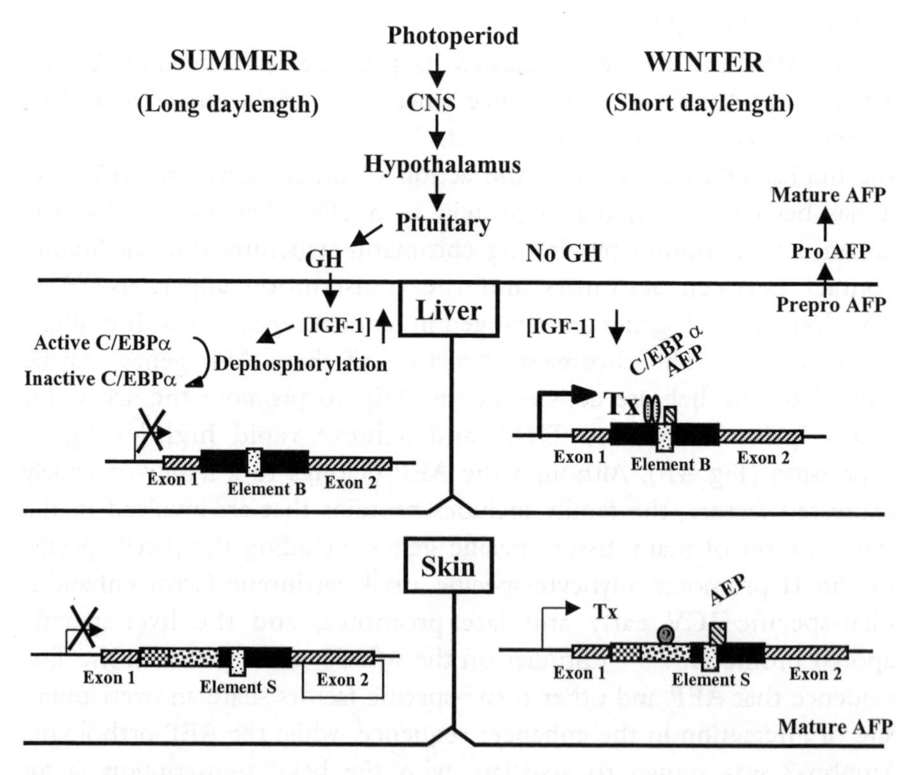

Fig. 5 Model for the hormonal and tissue-specific regulations of the wflAFP and wfsAFP genes.

transcriptional inhibition of the wflAFP genes. With the loss of long day length in the fall, the production of GH is inhibited by the CNS, thereby releasing the transcription repression. C/EBPα and AEP increase in concentration or activity and interact with the core enhancer Element B, which leads to wflAFP gene expression. The liver-enriched factor, C/EBPα, plays an important role not only in the seasonal and hormonal control, but also in the tissue-specificity of the wflAFPs. While lacking the C/EBP binding motif, the wfsAFP genes are more widely expressed and less influenced by GH and seasonal change. More studies are underway to investigate the role of AEP in the wfsAFP gene regulation and to determine whether there is a direct interaction between AEP and C/EBPα in wflAFP genes. The effect of seasonal and hormonal regulation on the structure and function of C/EBPα and AEP are also of interest and their study will lead to new knowledge on the complex mechanism of AFP gene regulation in winter flounder.

References

1. Harding MM, Ward LG and Haymet AD (1999). Type I "antifreeze" proteins. Structure-activity studies and mechanisms of ice growth inhibition. *Eur. J. Biochem.* **264**: 653–665.
2. Davies PL, Roach AH and Hew CL (1982). DNA sequence coding for an antifreeze protein precursor from winter flounder. *Proc. Natl. Acad. Sci. USA* **79**: 335–339.
3. Pickett M, Scott G, Davies P, Wang N, Joshi S and Hew C (1984). Sequence of an antifreeze protein precursor. *Eur. J. Biochem.* **143**: 35–38.
4. Hew CL, Wang NC, Yan S, Cai H, Sclater A and Fletcher GL (1986). Biosynthesis of antifreeze polypeptides in the winter flounder. Characterization and seasonal occurrence of precursor polypeptides. *Eur. J. Biochem.* **160**: 267–272.
5. Fourney RM, Joshi SB, Kao MH and Hew CL (1984). Heterogeneity of antifreeze polypeptides from the Newfoundland winter flounder, *Pseudopleuronectes americanus. Can. J. Zool.* **62**: 28–33.

6. Hew CL, Joshi S and Wang N-C (1984). Analysis of fish antifreeze polypeptides by reversed-phase high performance liquid chromatography. *J. Chromatogr.* **296**: 213–219.
7. Sicheri F and Yang DS (1995). Ice-binding structure and mechanism of an antifreeze protein from winter flounder. *Nature* **375**: 427–431.
8. Yang DS, Sax M, Chakrabartty A and Hew CL (1988). Crystal structure of an antifreeze polypeptide and its mechanistic implications. *Nature* **333**: 232–237.
9. Scott GK, Hew CL and Davies PL (1985). Antifreeze protein genes are tandemly linked and clustered in the genome of the winter flounder. *Proc. Natl. Acad. Sci. USA* **82**: 2613–2617.
10. Davies PL (1992). Conservation of antifreeze protein-encoding genes in tandem repeats. *Gene* **112**: 163–170.
11. Scott GK, Davies PL, Kao MH and Fletcher GL (1988). Differential amplification of antifreeze protein genes in the pleuronectinae. *J. Mol. Evol.* **27**: 29–35.
12. Gong Z, Fletcher GL and Hew CL (1992). Tissue distribution of fish antifreeze protein mRNAs. *Can. J. Zool.* **70**: 810–814.
13. Gong Z, Ewart KV, Hu Z, Fletcher GL and Hew CL (1996). Skin antifreeze protein genes of the winter flounder, *Pleuronectes americanus*, encode distinct and active polypeptides without the secretory signal and prosequences. *J. Biol. Chem.* **271**: 4106–4112.
14. Davies PL and Gauthier SY (1992). Antifreeze protein pseudogenes. *Gene* **112**: 171–178.
15. Slaughter, D and Hew, CL (1982). Radioimmunoassay for the antifreeze polypeptides of the winter flounder (*Pseudopleuronectes americanus*): seasonal profile and immunological cross-reactivity with other fish antifreezes. *Can. J. Biochem.* **60**: 824–829.
16. Chan SL, Fletcher GL and Hew CL (1993). Control of antifreeze protein gene expression in winter flounder. In: Hochachka and Mommsen (eds.), *Biochemistry and Molecular Biology of Fishes.* Elsevier Science Publishers, pp. 293–305.
17. Pickett MH, Hew CL and Davies PL (1983). Seasonal variation in the level of antifreeze protein mRNA from the winter flounder. *Biochim. Biophys. Acta* **739**: 97–104.

18. Fourney RM, Fletcher GL and Hew CL (1984). The effects of long day length on liver antifreeze messenger RNA in the winter flounder, *Pseudopleuronectes americanus*. *Can. J. Zool.* **62**: 1456–1460.
19. Fletcher GL, Idler DR, Vaisius A and Hew CL (1989). Hormonal regulation of AFP gene expression in winter flounder. *Fish Physiol. Biochem.* 7: 387–393.
20. Price JL, Gourlie BB, Lin Y and Huang RCC (1986). Induction of winter flounder antifreeze protein messenger RNA at 4°C *in vivo* and *in vitro*. *Physiol. Zool.* **59**: 679–695.
21. Fletcher GL (1981). Effects of temperature and photoperiod on the plasma freezing point depression, Cl^- and protein antifreeze in winter flounder (*Pseudopleuronectes americanus*). *Can. J. Zool.* **59**: 193–201.
22. Vaisius A, Martin-Kearley J and Fletcher GL (1989). Antifreeze protein gene transcription in winter flounder is not responsive to temperature. *Cell. Mol. Biol.* **35**: 547–554.
23. Duncker BP, Koops MD, Walker VK and Davies PL (1995). Low temperature persistence of type I antifreeze protein is mediated by cold-specific mRNA stability. *FEBS Lett.* **377**: 185–188.
24. Hew CL and Fletcher GL (1979). The role of pituitary in regulating antifreeze protein synthesis in the winter flounder. *FEBS Lett.* **99**: 337–339.
25. Fletcher GL (1979). The effects of hypophysectomy and pituitary replacement on the plasma freezing point depression, chloride ion concentration, glucose and protein antifreeze in winter flounder (*Pseudopleuronectes americanus*). *Comp. Biochem. Physiol.* **63**: 535–537.
26. Fletcher GL, King MJ and Hew CL (1984). How does the brain control the pituitary release of antifreeze synthesis inhibitor? *Can. J. Zool.* **62**: 839–844.
27. Fourney RM, Fletcher GL and Hew CL (1984). Accumulation of winter flounder antifreeze messenger RNA after hypophysectomy. *Gen. Comp. Endocrinol.* **54**: 392–401.
28. Idler DR, Fletcher GL, Belkhode S, King MJ and Hwang SJ (1989). Regulation of antifreeze protein production in winter flounder: a unique function for growth hormone. *Gen. Comp. Endocrinol.* **74**: 327–334.
29. Daughaday WH and Rotwein P (1989). Insulin-like growth factors I and II. Peptide, messenger ribonucleic acid and gene structures, serum, and tissue concentrations. *Endocr. Rev.* **10**: 68–91.

30. Pankov YA (1999). Growth hormone and a partial mediator of its biological action, insulin-like growth factor I. *Biochemistry (Mosc)* **64**: 1–7.
31. Duguay SJ, Swanson P and Dickhoff WW (1994). Differential expression and hormonal regulation of alternatively spliced IGF-I mRNA transcripts in salmon. *J. Mol. Endocrinol.* **12**: 25–37.
32. Shamblott MJ, Cheng CM, Bolt D and Chen TT (1995). Appearance of insulin-like growth factor mRNA in the liver and pyloric ceca of a teleost in response to exogenous growth hormone. *Proc. Natl. Acad. Sci. USA* **92**: 6943–6946.
33. Shepherd BS, Sakamoto T, Nishioka RS, Richman NH, 3rd, Mori I, Madsen SS, Chen TT, Hirano T, Bern HA and Grau EG (1997). Somatotropic actions of the homologous growth hormone and prolactins in the euryhaline teleost, the tilapia, *Oreochromis mossambicus. Proc. Natl. Acad. Sci. USA* **94**: 2068–2072.
34. Blakesley VA, Scrimgeour A, Esposito D and Le Roith D (1996). Signaling via the insulin-like growth factor-I receptor: does it differ from insulin receptor signaling? *Cytokine Growth Factor Rev.* 7: 153–159.
35. Petley T, Graff K, Jiang W, Yang H and Florini J (1999). Variation among cell types in the signaling pathways by which IGF-I stimulates specific cellular responses. *Horm. Metab. Res.* **31**: 70–76.
36. Gong Z, King MJ, Fletcher GL and Hew CL (1995). The antifreeze protein genes of the winter flounder, *Pleuronectus americanus*, are differentially regulated in liver and non-liver tissues. *Biochem. Biophys. Res. Commun.* **206**: 387–392.
37. Pickett MH, White BN and Davies PL (1983). Evidence that translational control mechanisms operate to optimize antifreeze protein production in the winter flounder. *J. Biol. Chem.* **258**: 14762–14765.
38. Gong Z and Hew CL (1993) Promoter analysis of fish antifreeze protein genes. In: Hochachka and Mommsen (eds.), *Biochemistry and Molecular Biology of Fishes.* Elsevier Science Publishers. pp. 307–324.
39. Chan SL, Miao M, Fletcher GL and Hew CL (1997). The role of CCAAT/enhancer-binding protein alpha and a protein that binds to the activator-protein-1 site in the regulation of liver-specific expression of the winter flounder antifreeze protein gene. *Eur. J. Biochem.* **247**: 44–51.
40. Miao M, Chan SL, Hew CL and Fletcher GL (1998). Identification of nuclear proteins interacting with the liver-specific enhancer B element of

the antifreeze protein gene in winter flounder. *Mol. Mar. Biol. Biotechnol.* 7: 197–203.

41. Hew C, Poon R, Xiong F, Gauthier S, Shears M, King M, Davies P and Fletcher G (1999). Liver-specific and seasonal expression of transgenic Atlantic salmon harboring the winter flounder antifreeze protein gene. *Transgenic Res.* **8**: 405–414.
42. Takiguchi M (1998). The C/EBP family of transcription factors in the liver and other organs. *Int. J. Exp. Pathol.* **79**: 369–391.
43. Landschulz WH, Johnson PF, Adashi EY, Graves BJ and McKnight SL (1988). Isolation of a recombinant copy of the gene encoding C/EBP [published erratum appears in *Genes Dev* (1994) May 1; **8**(9): 1131]. *Genes Dev* **2**: 786–800.
44. Montell DJ, Rorth P and Spradling AC (1992). Slow border cells, a locus required for a developmentally regulated cell migration during oogenesis, encodes Drosophila C/EBP. *Cell* **71**: 51–62.
45. Miao M, Chan SL, Hew CL and Gong Z (1998). The skin-type antifreeze protein gene intron of the winter flounder is a ubiquitous enhancer lacking a functional C/EBP alpha binding motif. *FEBS Lett.* **426**: 121–125.
46. Hemati N, Ross SE, Erickson RL, Groblewski GE and MacDougald OA (1997). Signaling pathways through which insulin regulates CCAAT/ enhancer binding protein alpha (C/EBP alpha) phosphorylation and gene expression in 3T3-L1 adipocytes. Correlation with GLUT4 gene expression. *J. Biol. Chem.* **272**: 25913–25919.
47. Karin M, Liu Z and Zandi E (1997). AP-1 function and regulation. *Curr. Opin. Cell. Biol.* **9**: 240–246.
48. Crossley M and Brownlee GG (1990). Disruption of a C/EBP binding site in the factor IX promoter is associated with haemophilia B. *Nature* **345**: 444–446.
49. Miao M, Chan SL, Fletcher GL and Hew CL (2000). The rat ortholog of the presumptive flounder antifreeze enhancer-binding protein is a helicase domain-containing protein. *Eur. J. Biochem.* **267**: 7237–7246.
50. Shieh SY, Stellrecht CM and Tsai MJ (1995). Molecular characterization of the rat insulin enhancer-binding complex 3b2. Cloning of a binding factor with putative helicase motifs. *J. Biol. Chem.* **270**: 21503–21508.
51. Fukita Y, Mizuta TR, Shirozu M, Ozawa K, Shimizu A and Honjo T (1993). The human S(mu)bp-2, a DNA-binding protein specific to the

single-stranded guanine-rich sequence related to the immunoglobulin mu chain switch region. *J. Biol. Chem.* **268**: 17463–17470.

52. Mizuta TR, Fukita Y, Miyoshi T, Shimizu A and Honjo T (1993). Isolation of cDNA encoding a binding protein specific to 5'-phosphorylated single-stranded DNA with G-rich sequences. *Nucleic Acids Res.* **21**: 1761–1766.
53. Kerr D and Khalili K (1991). A recombinant cDNA derived from human brain encodes a DNA binding protein that stimulates transcription of the human neurotropic virus JCV. *J. Biol. Chem.* **266**: 15876–15881.
54. Eisen A and Lucchesi JC (1998). Unraveling the role of helicases in transcription. *Bioessays.* **20**: 634–641.
55. Laurent BC, Yang X and Carlson M (1992). An essential *Saccharomyces cerevisiae* gene homologous to SNF2 encodes a helicase-related protein in a new family. *Mol. Cell. Biol.* **12**: 1893–1902.
56. Sebastiani G, Durocher D, Gros P, Nemer M and Malo D (1995). Localization of the Catfl transcription factor gene to mouse chromosome 19. *Mamm. Genome* **6**: 147–148.
57. Chen NN, Kerr D, Chang CF, Honjo T and Khalili K (1997). Evidence for regulation of transcription and replication of the human neurotropic virus JCV genome by the human S(mu)bp-2 protein in glial cells. *Gene* **185**: 55–62.
58. Mohan WS, Chen ZQ, Zhang X, Khalili K, Honjo T, Deeley RG and Tam SP (1998). Human S(mu) binding protein-2 binds to the drug response element and transactivates the human apoA-I promoter: role of gemfibrozil. *J. Lipid Res.* **39**: 255–267.
59. Zhang Q, Wang YC and Montalvo EA (1999). Smubp-2 represses the Epstein-Barr virus lytic switch promoter. *Virology* **255**: 160–170.

Chapter 7

The Skin-Type Antifreeze Polypeptides: A New Class of Type I AFPs

Woon-Kai Low
Department of Biochemistry
University of Toronto
Toronto, Ontario, M5G 1L5, Canada

Qingsong Lin*
†*Department of Laboratory Medicine and Pathobiology*
University of Toronto
Toronto, Ontario, M5G 1L5, Canada

K Vanya Ewart
NRC Institute for Marine Biosciences
Halifax, Nova Scotia, B3H 3Z1, Canada

Garth L Fletcher
Ocean Sciences Center
Memorial University of Newfoundland
St. John's, Newfoundland, A1C 5S7, Canada

Choy L Hew*†
Tropical Marine Science Institute
National University of Singapore
Singapore 117543

Introduction

Antifreeze proteins (AFPs) and antifreeze glycoproteins (AFGPs) are produced by a variety of fish species that survive in icy seawaters. Most of the earlier characterized AF(G)Ps were isolated from fish sera and are produced mainly by the liver. A few early studies examined for the presence of AF(G)Ps in other tissues. The presence of AFGPs in the intestinal[1] and interstitial fluids of all body tissues except the brain of Antarctic notothenioid fish was reported,[2] and type I AFPs were isolated from skin extracts of Arctic-boreal sculpin (*Myoxocephalus scorpius*).[3] In addition, thermal hysteresis activity was detected in skin extracts from the cunner (*Tautogolabrus adspersus*).[4] However, the investigation of non-serum antifreeze proteins took a new turn when AFP-related mRNA was found in many non-liver tissues of the winter flounder (*Pleuronectes americanus*).[5] This study also reported the presence of type III AFP mRNA in the skin, dorsal fin, gills, intestine and stomach of the ocean pout. In 1996, Gong *et al.*[6] isolated and cloned the non-serum AFPs from winter flounder and suggested that they were produced as mature polypeptides within the cytosol due to the lack of signal peptides and prosequences in the mRNA sequences. This new class of type I AFPs was termed skin-type AFP because they were originally cloned and isolated from winter flounder skin, even though they are ubiquitously expressed in that species. The term skin-type AFP is now used to describe the antifreeze proteins that lack signal sequences and have to date been isolated and characterized in three different species of fish. The lack of signal and prosequences in these AFPs implies that skin-type AFPs can be found as mature intracellular polypeptides. Using the nomenclature system proposed in Low *et al.*,[7] the serum AFPs were given the name liver-type AFP to denote their predominant site of production in the liver. Thus, for winter flounder, the skin-type AFPs are referred to as wfsAFPs, and the liver-type AFPs are referred to wflAFPs. The discovery of the skin-type AFPs may reveal new dimensions of biological freezing protection provided by AFPs, whose function in fish was previously defined solely as a means of freezing point depression in extracellular body fluids. This chapter will review the isolation and cloning of the skin-type AFPs, highlight some of the structural/functional studies on skin-type AFPs, and speculate on the possible biological roles for an intracellular antifreeze protein.

Identification of Skin-Type AFPs

Winter Flounder Skin-Type AFPs

The winter flounder produces type I AFPs, which are synthesized primarily in the liver as preproAFPs[8,9] (see Fig. 2), that are processed and secreted into the circulation. The most abundant winter flounder AFP is wflAFP-6 (HPLC-6). WflAFP-6 has been viewed as the prototypical type I AFP and is the most widely studied (reviewed in Refs. 10–13). More recently, Gong *et al.* (1992)[5] found that a substantial amount of AFP mRNA was present in many other tissues of the winter flounder, notably the skin, scales and gills. Furthermore, the expression of AFPs from non-liver tissues is regulated differently from the liver-type AFPs. The expression of skin-type AFPs exhibits only a modest (five- to ten-fold) seasonal variation compared with the approximately 700-fold variation of the liver-type AFPs and their transcription is not regulated by growth hormone in contrast to their serum counterparts.[14]

Screening of a winter flounder skin cDNA library with a liver-type AFP cDNA produced 14 clones. Nine distinct DNA sequences encoding eight AFPs, wfsAFP-1 to -8 (previously named as sAFP1 to sAFP8) were characterized[6] (Fig. 1). All of the clones, except for P13 and S6, which encode additional C-terminal sequences, share a higher sequence identity to each other (91.7–99.2%) than to the wflAFP DNA sequences (72.1–82.2%). The polypeptides encoded by the skin-type clones nonetheless have approximately 45–55% Ala content, and contain the basic 11-residue repeat motif that has been traditionally described as $T\text{-}X_2\text{-}D/N\text{-}X_7$ or $T\text{-}X_{10}$ (where X is predominantly Ala) (see below). Two sets of AFP genes, skin-type and liver-type AFP were defined, reflecting the differences in DNA sequence between the skin-type and liver-type AFP cDNA clones. Comparisons among the wfsAFPs show them to be almost identical and the few amino acid differences are restricted to the last few residues at the C-terminal end (Fig. 1). Unlike the liver-type AFP genes, which encode the secreted AFPs, the skin-type AFP genes encode mature polypeptides without the pre- and prosequences, consistent with an intracellular localization (Fig. 2).

```
                                         11-residue repeat motif
                                         TaaxAxxAAxx TaaxAxxAAxx
                              1          11          22          33
wfsAFP-1 (S4)                 MDAPARAAAA TAAAAKAAAEA TKAAAAKAAAA TKAAAH
wfsAFP-2 (P12, S3)            MDAPAKAAAA TAAAAKAAAEA TAAAAAKAAAA TKAGAAR
wfsAFP-3 (P9)                 MDAPAKAAAA TAAAAKAAAEA TAAAAAKAAAD TKAKAAR
wfsAFP-4 (P7, P8)             MDAPAKAAAA TAAAAKAAAEA TAAAAAKAAAA TKAGAAH
wfsAFP-5 (L3, S2, S11)        MDAPAKAAAA TAAAAKAAAEA TAAAAKAAAAA TKAAA
wfsAFP-6 (L4)                 MDAPAKAAAA TAAAAKAAAEA TAAAAAKAAAA *KAAGHAR
wfsAFP-7 (S9, S10)            MDAPAAAAAA TAAAAKAAAEA TAAAAAKAAAA TKAAAAR
wfsAFP-8 (P13, S6)            MDAPAAAAAA TAAAAKAAAEA TAAAAAAAAAA TAEAAKAAAATKAAAAAAAAR
F2                            MDAPAKAAAA TAAAAKAAAEA TAAAAAKAAAA TKAGAAR
11-3                          MDAPAKAAAA TAAAAKAAAEA TAAAAAKAAAA TKAAAHAR

lssAFP(clone)                 MDAPAKAAAK TAADAKAAAAK TAADALAAANK TAAAAKAAAK
lssAFP-8                      MDAPAKAAAK TAADAKAAAAK TAADALAAANK TAAAAKAAA

sslAFP-3                      MDAPARAAAK TAADALAAANK TAADAAAAAAA A
gslAFP-5                      MDAPAIAAAK TAADALAAAKK TAADAAAAAAK P

wflAFP-6                  DTASDAAAAAAL TAANAKAAAEL TAANAAAAAAA TAR

sssAFP-2    MAAAAKAAEA AAMAAANAAE AAATKAADAA ASAAAAAIAA IAEAAEAAEA AATKSANVAA
            AAAATSAAAA AKATANAAAA ASAAAAAAAA VA
```

Fig. 1 A comparison of type I AFP sequences. The amino acid sequences of wfsAFP-1 to -8 were deduced from cDNA clones (listed in parentheses); F2 and 11–3 are from winter flounder genomic clones; some liver-type AFPs are also included for comparison. Conserved N-terminal MDAPA sequence (red); conserved 11-residue motif Thr (blue); completely conserved positions (grey highlight); conserved positions within wfsAFPs (green); and conserved positions within sculpin AFPs (pink). The sequence of sssAFP-2 is not aligned with the other type I AFPs.

```
                       10         20  ▽      30         40    ↓     50         60         70         80
preprowflAFP-6  MALSLFTVGN LIFLFWTMRI TEASPDPAAK AAPAAAAAPA AAAPDTASDA AAAAALTAAN AKAAAELTAA NAAAAAAATA R
preprowflAFP-8  MALSLFTVGN LIFLFWTMRI TEASPDPAAK AAPAAAAAPA AAAPDTASDA AAAAALTAAN AAAAAKLTAD NAAAAAAATA R
wfsAFP-1                                                          MDAP ARAAAATAAA AKAAAEATKA AAAKAAAATK AAAH
wfsAFP-2                                                          MDAP AKAAAATAAA AKAAAEATAA AAAKAAAATK AGAAR
wfsAFP-3                                                          MDAP AKAAAATAAA AKAAAEATAA AAAKAAADTK AKAAR
```

Fig. 2 Skin-type AFPs are produced as mature polypeptides lacking the pre- and prosequences found in serum AFPs. A comparison of the wflAFP-6 and -8 sequences deduced from cDNA sequences that show the pre- and prosequences with the sequences of wfsAFP-1 to -3. The triangle displays the junction between the pre- and prosequence, and the arrow indicates the junction between the proAFP and mature polypeptide.[8,9]

A comparison of the skin-type AFP cDNA sequences revealed that they were closely related to the genomic sequences, F2 and 11–3.[15] These two genes were initially thought to be pseudogenes due to the presence of in-frame stop codons in the 5′-upstream region that would have corresponded to the prosequence in the liver-type AFPs and the lack of a TATA box in the putative promoter region. However, F2 encodes a polypeptide sequence that is identical to the wfsAFP-2. Similarly, the polypeptide sequence encoded by 11–3 differs by only one amino acid from wfsAFP-2 sequence. Therefore, both F2 and 11–3 code for functional AFPs in skin and other peripheral tissues. Primer extension was carried out to confirm that the isolated clones did in fact contain full-length coding sequences and to map the transcription start site of the skin-type AFP genes. A putative TFIID binding motif AATAAAT was found 25 nucleotides upstream of the first start site, further indicating that F2 and 11–3 are functional. These observations further supported the evidence that the skin-type AFPs are produced without signal or prosequences.

In the same study,[6] reverse-phase HPLC of winter flounder skin scrapings identified three new Ala-rich antifreeze polypeptides that were distinct from the major liver-type AFPs, wflAFP-6 and -8. The polypeptides were named wfsAFP-1 to -3 and were correlated with the corresponding cDNA and gene sequences (see Fig. 1). An additional 42 mass units observed in the purified polypeptides indicated that these AFPs were N-terminally blocked with an acetyl group.

Skin-Type AFPs in Other Species of Fish

Following the discovery of the wfsAFPs, other species of fish were examined for skin-type AFPs. The goal was to determine whether skin-type AFPs are unique to the winter flounder or if they are a more universal form of type I AFP. To begin this comparative study, the shorthorn sculpin (*Myoxocephalus scorpius*) was chosen because it also produces serum type I AFPs,[16–18] and previous work had suggested the possible existence of AFPs in the skin.[3] Furthermore, the winter flounder and shorthorn sculpin occupy the same waters off the coast of Newfoundland and would experience similar environmental conditions. If skin-type AFPs are an essential component in freezing protection, it would be reasonable to expect the shorthorn sculpin to produce a counterpart to the winter flounder skin-type AFPs. A second fish species, the longhorn sculpin (*Myoxocephalus octodecemspinosus*) was also investigated because of its close relation to the shorthorn sculpin.[19] Interestingly, despite the close evolutionary relationship between the longhorn and shorthorn sculpin, the longhorn sculpin does not appear to produce serum type I AFPs. It produces the unique type IV AFP, albeit at much lower concentrations than those of serum antifreeze proteins in other species.[20–22] This is consistent with the apparent lower requirement for AFPs in longhorn sculpin, which is found in relatively deeper waters than the shorthorn sculpin and winter flounder and migrates offshore in the winter season, thereby avoiding encounters with environmental ice.

Shorthorn Sculpin

Screening of a cDNA library of shorthorn sculpin skin tissue, using a wfsAFP clone as a probe, produced two cDNA sequences that were later determined to be essentially identical, with one cDNA containing more of the 5′-untranslated region.[7] This cDNA sequence contained a single ORF defined by an ATG start codon and TAG stop codon that encoded an Ala-rich polypeptide. The ORF also lacked a signal peptide, and prosequence, suggesting an intracellular localization. Southern blotting experiments revealed a multigene family for the shorthorn sculpin skin-type AFPs (unpublished results). Notable characteristics of this polypeptide are that it is extremely Ala-rich (~70%) even for a type I AFP which generally have 50–60% alanine. Second, this 92-residue AFP is the largest naturally occurring type I AFP identified to date. Third, it lacks a defined 11-residue repeat motif structure (see below). The production of a recombinant protein in *E. coli* revealed that the polypeptide sequence encoded by the clone is a highly active type I AFP with activity levels comparable to wflAFP-6. The polypeptide was named sssAFP-2 under the new nomenclature system (Fig. 1).[7]

Examination of the tissue expression pattern by Northern blotting, RT-PCR and primer extension analysis, showed sssAFP-2 to be predominantly expressed in peripheral tissues, similar to the winter

Table 1 Tissue expression patterns of the skin-type AFPs of the winter flounder compared with the shorthorn and longhorn sculpins. Skin-type AFPs are predominantly found in the peripheral tissues such as skin and gills. Level of relative mRNA production is indicated by number of (+). The (+/–) indicates very low, but detectable, levels of mRNA.

	Winter flounder			
Tissue	Skin-type AFPs	Liver-type AFPs	Shorthorn Sculpin	Longhorn Sculpin
Liver	+/–	+++++	–	+/–
Stomach	++++	–	++++++	++
Intestine	+	–	N/A	++
Kidney	+	–	–	++
Spleen	+/–	–	N/A	N/A
Gills (filaments)	++++++	–	++++++	+++++
Fin	++++++	–	++++++	++++++
Scales	++++++	–	N/A	N/A
Skin	++++++	–	++++++	++++++
Muscle	N/A	N/A	–	+/–
Brain	N/A	N/A	++++	+/–
Methods of detection	Northern	Northern	Northern RT-PCR Primer extension	Northern RT-PCR

flounder skin-type AFPs (Table 1). All three methods showed the expression of sssAFP-2 in brain, which is an unusual location for AFPs. A notable difference between the skin AFPs of sculpin and flounder is that expression of sssAFP-2 was undetectable in the liver by all three methods, in contrast to the winter flounder, where it is ubiquitously expressed. Furthermore, whereas the wfsAFPs show only a moderate (five- to ten-fold) seasonal variation in expression levels, a dramatic seasonal variation exists for sssAFP-2, ranging from trace levels in summer to very strong signals in winter as evidenced by Northern blotting.

Longhorn Sculpin

Northern blot analysis of longhorn sculpin total RNA tissues using the 3′-UTR of the sssAFP-2 clone as a probe produced specific bands in the skin, gill filament and dorsal fin, but not in the liver. A type I ORF sequence that was 42 residues long, Ala-rich, and contained an 11-residue repeat structure was eventually cloned using RT-PCR and RACE methods.[23] The corresponding polypeptide was designated as lssAFP and four individual Ala-rich polypeptides with antifreeze activity were isolated from the skin tissue. One of these polypeptides (lssAFP-8) has been fully sequenced using tandem mass spectrometry[23] and only differs from the clone sequence in lacking the C-terminal Lys residue. Furthermore, a recombinant protein based on the first 38 residues of the clone sequence was produced and its antifreeze activity was found to be comparable to that of wfsAFP-2 (Low *et al.*, manuscript in preparation). From Southern analysis, it was found that the lssAFPs are encoded by a multigene family, which is a consistent feature of AFPs. RT-PCR and immunoblotting analysis indicate that the highest expression of skin-type AFPs in the longhorn sculpin occurs in the peripheral tissues, most notably the skin and gills, with lower levels detected in the intestinal tissues such as the stomach.

Evolutionary Relationships Among Skin-Type AFPs

Among teleost fish, at least five distinct biochemical classes of proteins have evolved a similar function of ice-binding and thermal hysteresis. Type I AFPs can be found in many species, although no progenitor gene has been identified. Indeed, the sculpins and flounders are only related at the superorder level (Acanthopterygii).[19] Direct comparisons of the polypeptide sequences between type I AFPs are further complicated by the fact that they all possess 40–60% Ala. Thus, comparisons always provide high identity and may therefore provide misleading information. However, sequence comparisons of type I AFPs should not be so quickly dismissed as insights into their relationships can be gained when specific criteria are used. For example, when the type I AFPs are grouped according to their 11-residue repeat content, the longhorn sculpin skin-type AFP appears more closely related to the winter flounder skin-type AFPs than to the shorthorn sculpin skin-type AFP (Fig. 1). The next section will briefly discuss some of the relationships found amongst the different skin-type AFPs as well as those with other type I AFPs.

Comparison of the skin-type AFP amino acid sequences with other known type I serum AFPs (Fig. 3A), shows that lssAFP shares the highest % sequence identity with the minor sculpin serum AFPs. The next highest % of sequence identity is with the wfsAFPs (represented by wfsAFP-3). This relationship arises from the similar 11-residue repeat structure and the N-terminal MDAPA sequence. Surprisingly, sssAFP-2 is the least similar to lssAFP, even though their respective clone sequences are nearly identical in their untranslated regions.[23] This is illustrated in Fig. 3B, where the 5′-UTRs of four type I AFP gene sequences are compared. This high % identity also exists in the 3′-UTR and the highest region of variability between sssAFP-2 and lssAFP lies in the ORFs.

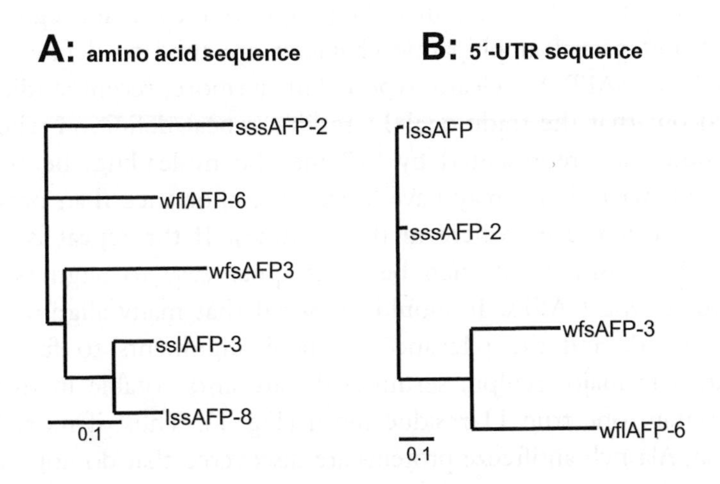

Fig. 3 Comparisons of the primary sequences **(A)** and 5′-UTR sequences **(B)** of several type I AFPs. Sequence comparisons were performed using ClustalX (1.8) to create unrooted neighbor-joining trees which were visualized using TreeView (Win32) (1.6.1).[57–59] The sslAFP-3 is not included in the 3′ UTR tree because no cDNA is available for it to date.

When the ORFs of wfsAFP-6 and wfsAFP-3 are compared, their close relationship is apparent. However, lssAFP and sssAFP are now much more similar to one another than in the amino acid sequence comparison. This new relationship reflects differences in codon usage between the flounder and sculpins (see Fig. 4). In *Drosophila*, which has served as a model multicellular eukaryote for codon usage bias analysis, most genes are found to conserve their codon usage bias.[24] Thus, the lack of conserved codon usage in the flounder and sculpins would tend to suggest that the two species have evolutionarily unrelated AFP genes.

It is evident that the longhorn sculpin and shorthorn sculpin skin-type AFPs are homologous genes from the level of sequence identity between their cDNA UTR regions. Nonetheless, it could be argued that the shorthorn sculpin skin-type AFP is not a "true" type I AFP because its polypeptide sequence diverges too much from the more predominant forms of type I AFP. The differences are its larger size, and lack of recognizable 11-residue Thr repeats. However, if general structural features, including high Ala content, high helical content and partial amphipathic character are used to defined type I AFPs, then sssAFP-2 is clearly type I. Furthermore, recent studies have pointed out that the traditional 11-residue repeat definition where the Ala residues are represented by "x" may be misleading, because the conserved Ala residues may have a greater importance than previously believed (reviewed elsewhere in this volume). If the repeat is defined by the Ala positions, it then becomes quite easy to align sssAFP-2 with other type I AFPs. It should be noted that many alignments can be generated, and the relevance of these alignments to function is unclear. The major sculpin serum AFPs are also notable in that they only contain one true 11-residue motif (Fig. 1). Thus, if more highly α-helical, Ala-rich antifreeze proteins are discovered that do not conform to the traditional 11-residue repeat structure, it may become necessary to divide the type I AFP into structural subclasses.

It is still unclear if the sculpin AFPs share an evolutionary relationship or a common ancestor gene with flounder AFPs. The flounder skin-type AFP genes were initially described as pseudogenes that were

A: ORF sequence

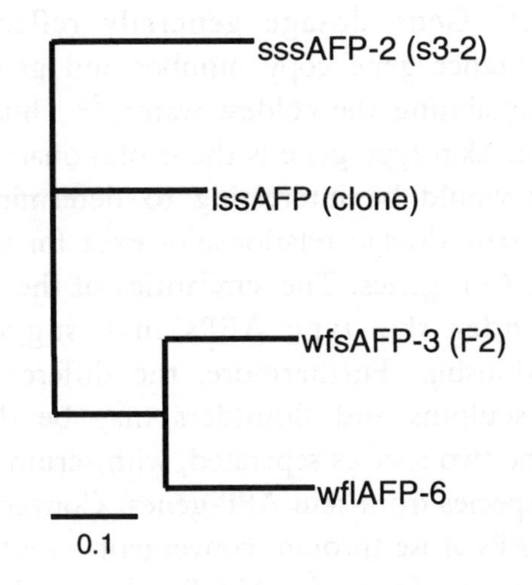

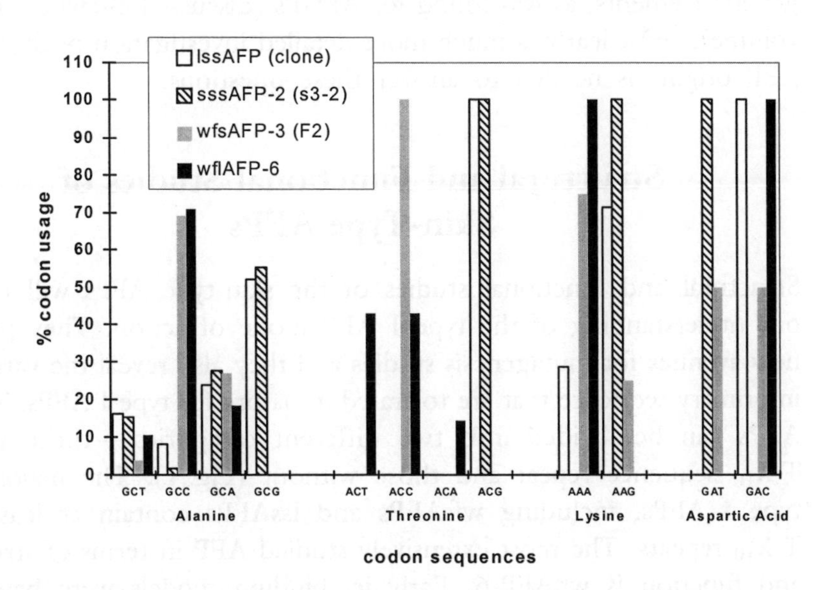

Fig. 4 The winter flounder and sculpin skin-type AFP genes show differences in preferred codon usage. **(A)** Unrooted neighbour-joining tree comparison of the ORF sequences for various type I AFPs. **(B)** The percent codon usage for four typical type I AFP amino acids in different type I AFPs.

considered as possible progenitor genes for the liver-type AFPs[15] that duplicated and amplified in copy number to form tandemly repeated liver-type genes.[25] Gene dosage generally reflects evolutionary adaptation, with higher gene copy number and gene dosage in the flounder species inhabiting the coldest waters.[26] Thus, it appears that in the flounder, the skin-type gene is the evolutionary precursor of the liver-type, and it would be interesting to determine whether such evolutionary and gene dosage relationships exist for the longhorn and shorthorn sculpin AFP genes. The similarities of the longhorn sculpin and winter flounder skin-type AFPs may suggest a divergent evolutionary relationship. Furthermore, the differences between the serum AFPs of sculpins and flounders may be due to sequence divergence after the two species separated, with serum (liver-type) AFPs evolving in both species from skin AFP genes. Conversely, it is possible that liver-type I AFPs arose through convergent evolution from distinct genetic elements, as was found for AFGPs (discussed elsewhere in this volume).[27,28] Clearly, a much more detailed investigation of the type I AFP origins is needed to answer these questions.

Structural and Functional Studies of Skin-Type AFPs

Structural and functional studies of the skin-type AFPs will expand our understanding of the type I AFP mode of action. They provide new avenues for mutagenesis studies and they also reveal the variations in primary sequence that are tolerated in nature for type I AFPs. Type I AFPs can be divided into two different categories — those with a T-X_{10} sequence repeat and those without (Fig. 1). The majority of type I AFPs, including wfsAFPs and lssAFPs contain at least two T-X_{10} repeats. The most extensively studied AFP in terms of structure and function is wflAFP-6. Early ice-binding models were based on the assumption that hydrogen bonding was the major interaction for ice-binding.[27,28] More recent models suggest that the roles of van der Waals interactions and/or hydrophobic interactions in ice-binding are more significant.[13,29–31] While information gained from structure/

function studies on wflAFP-6 can be applied to some type I AFPs, for others, the structural diversity prevents this extrapolation. For example, sslAFP-8 has been shown to bind to a different plane of the ice crystal than wflAFP-6, and it has only one T-X_{10} motif. Knight *et al.* (1991)[28] have proposed a model that emphasizes the importance of charged amino acids in sslAFP-8 ice-binding and Wierzbicki *et al.* (1996)[32] have also inferred that charged/polar residues are involved based on modeling studies. However, there is no experimental support for these models. As sssAFP-2 lacks the traditional T-X_{10} repeats, it also might have a more unique mechanism of action.

Sequence alignment (Fig. 1) shows that wfsAFPs have a conserved N-terminal sequence, MDAPA, which is distinct from wflAFP N-terminal sequence DTASD. This MDAPA sequence is also found in sculpin serum AFPs and the lssAFPs. As well, the MDAPA-containing AFPs that have been characterized so far all have N-terminal acetylation. However, the role of N-terminal acetylation in native wfsAFPs is insignificant, at least in terms of antifreeze activity, as the bacterial expressed wfsAFP-2 lacking such modification possesses essentially the same level of activity as the native polypeptide.[33] The N-terminal region of the wflAFP-6 has been proposed to be an important feature for maintaining the high helical content of the polypeptide as this region possesses an elaborate hydrogen-bonding network forming an N-terminal "cap" structure.[34] The MDAPA sequence would likely have a different structure or N-terminal hydrogen bonding network. Whether the N-terminal sequence of skin-type AFPs is important for antifreeze activity because of structural characteristics or because of other biological functions remains to be determined. Interestingly, the MDAPA-containing AFPs have much lower thermal hysteresis activity than wflAFP-6. Thus, an examination of the structural differences may shed new light on their mechanism of ice-binding.

Based on the X-ray crystal structure of wflAFP-6, Sicheri and Yang (1995)[34] postulated that some amino acid residues, "-T--D-" and "-LT--N-", form relatively flat surfaces, ice-binding motifs (IBMs), to strengthen ice-binding. The wfsAFPs generally lack such IBMs, and thus, it was suggested that this might explain the difference in antifreeze

```
          1          11          22          33
wfsAFP-2  MDAPAKAAAA TAAAAKAAAEA TAAAAAKAAAA TKAGAAR
wfsAFP-3  MDAPAKAAAA TAAAAKAAAEA TAAAAAKAAAD TKAKAAR
0IBM      MDAPAKAAAA TAAAAKAAAEA TAAAAAKAAAA TKAAAAR
1IBM      MDAPAKAAAA TAAAAKAAAEA TAAAAAKAAAD TKAKAAR
2IBM      MDAPAKAAAA TAAAAKAAAED TAAKAAKAAAD TKAKAAR
3IBM      MDAPAKAAAD TAAKAKAAAED TAAKAAKAAAD TKAKAAR
Mut1      MDAPAKAAAD TAAAAKAAAEA TAAAAAKAAAA TKAGAAR
Mut2      MDAPAKAAAA TAAAAKAAAEA TAAAAAKAAAD TKAGAAR
Mut3      MDAPAKAAAD TAAAAKAAAEA TAAAAAKAAAD TKAGAAR
```

Fig. 5 Sequences of experimental wfsAFP mutants and analogues. 0 to 3IBM are synthetic peptides containing 0–3 putative IBM "-DT--K-"; Mut1 to 3 contain the putative IBM "-K---DT-". The putative ice-binding residues are highlighted.

activity between these two types of flounder AFPs. However, wfsAFP-3 does contain several residues that might meet the IBM criteria. There are two possibilities, one involving D32, T33 and K36, the other involves K28, D32 and T33. To test the first putative IBM "-DT--K-", a series of synthetic peptides were generated, which contain 0, 1, 2 or 3 putative IBMs.[34] To examine the other possibility, "-K---DT-" was introduced by site-directed mutagenesis into the N- or C-terminal region, or both, of wfsAFP-2 which does not contain such IBMs.[33] The amino acid sequences of these mutants are shown in Fig. 5. In both cases, the IBMs did not enhance the antifreeze activity. The activity actually decreased as greater numbers of putative IBMs were introduced and there was a corresponding decrease in the helix stability of the AFPs. Thus the roles of "-K---DT-" or "-DT--K-" as functional IBMs were not substantiated. However, the introduction of the wflAFP-6 IBM "-LT--N-" into a skin-type AFP, resulted in ~15% activity enhancement (Lin and Hew, unpublished data) which provides experimental support for the IBM hypothesis. However, the dramatic difference in activity between the skin- and liver-type AFPs cannot be explained based solely on the nature of the IBMs alone. Other structural differences may be involved. The bacterially expressed wflAFP-6, which lacked C-terminal amidation, had only ~70% helix at 0°C and maintained only approximately 70% activity of the C-amidated native polypeptide, which is essentially 100% helical at 0°C[36] (Lin, Yang and Hew, unpublished data). Thus, the lower activities of wfsAFPs might also be partly due to the lack of C-terminal amidation (see below).

In the type I AFPs, a correlation exists between antifreeze activity and helix stability. The introduction of two additional salt bridges into wflAFP-6 enhanced its helix content,[37] and although this highly helical analogue displayed almost identical antifreeze activity as the native AFP, it was able to affect the ice crystal growth rates at seven- to eight-fold lower concentrations. The correlation of helix content with antifreeze activity was further confirmed in studies by Wen and Laursen (1993).[38] More recently, Houston *et al.* (1998)[39] have shown that the introduction of an internal lactam bridge into a minimized AFP analogue that only contains one T-X_{10} repeat reinforced the helix content of this peptide and conferred ice-binding ability upon it. In the IBM studies mentioned above, the decrease in activity correlated with the decrease of AFP helix content, which might be due to replacing helix favorable Ala with less favorable Asp. As well, wflAFP-6 only has 70% activity and helix content when lacking C-terminal amidation. C-terminal amidation stabilizes the helix by participating in the formation of cap structure or neutralizing helix dipole.[34] Type I AFPs are believed to form rigid helices in order to match their ice-binding residues to the ice lattice. Higher helix content in solution might mean that less energy is required for a conformational transition when the AFP binds to ice, thus conferring an apparent higher antifreeze activity.

According to the adsorption-inhibition model,[28,40] specific hydrophilic groups on the AFP hydrogen bond with atoms on a specific plane of the ice lattice. Knight *et al.*[27] have hypothesized that some hydrogen bonding groups on the AFPs might become incorporated into the ice lattice so that they can each form three hydrogen bonds. More recently, based on the X-ray structure, Sicheri and Yang (1995)[33] suggested that the ice-binding groups are too close to the helix backbone to protrude far enough into the ice to become tetrahedrally linked. However, the importance of hydrogen bonding in AFP-binding to ice has recently been challenged. Two groups have independently reported that the replacement of the Thr residue with Val maintains antifreeze freeze activity.[29–31] Chao *et al.* (1998)[29] first reported that the replacement of the Thr residues with Val in the central two repeats of wflAFP-6 largely maintained the antifreeze activity. Furthermore, replacing the two Thr residues with Ser, which would maintain the hydrogen bonding capacity at these positions, resulted in significant loss of antifreeze activity. This suggested a diminished role of hydrogen bonds in AFP-binding to ice. Haym *et al.* (1998 and 1999)[30,31] essentially confirmed these results by replacing all four Thr residues. However, there are some discrepancies in terms of the exact role of hydrogen bonds between the various studies. Zhang and Laursen (1998)[41] reported that the replacement of all four Thr residues of wflAFP-6 to Val residues resulted in significant decrease of antifreeze activity (about 31% of native activity) and suggested that hydrogen bonds still play an important role. However, Haymet *et al.*[30] reported the same Thr→Val mutations maintained 100% of native activity and claimed that hydrogen bonds had no role in ice-binding, and hydrophobic interaction was the major driving force. In order to investigate these discrepancies, we have decided to use wfsAFP-2 as a model to study the effects of Thr replacements on antifreeze activity. WfsAFP-2 has an advantage over wflAFP-6 in that the three Thr residues of wfsAFP-2 are not likely to be involved in the cap structures and thus will contribute more equally to ice-binding. In addition, wfsAFP-2 has more positively charged residues, thus mutating Thr residues to hydrophobic residues should not result in solubility problems, as found in wflAFP-6. Our results have shown that replacing the three Thr of wfsAFP-2 with Val maintains 100% activity, thus supporting the notion that hydrogen bonding by the hydroxyls of Thr is not part of the type I AFP mechanism of ice-binding (Lin and Hew, unpublished results).

The 11-residue repeat motifs in wflAFPs have been traditionally defined as T-X_2-D/N-X_7 (where X is any amino acid, but usually Ala). It is generally understood that the positions indicated by X are usually Ala, but a comparison of most of the known type I AFP sequences indicates that there are positions that are always Ala (Fig. 1). Moreover, the traditional definition may place too much emphasis on the residue in the 4th position, i.e. the wfsAFPs are all Ala at this position. Considering flounder and sculpin skin-type AFPs, as well as some minor sculpin serum AFPs (Fig. 1), the repeat motif may be re-defined as TaaxAxxAAxx (where A is mainly Ala, and x can be any amino acid including Ala) (Fig. 1). Thus, most type I AFPs have a highly conserved hydrophobic face (see Fig. 6). The Thr may also be involved in this face with the γ-methyl group oriented towards it.[31,34] As the perceived role of hydrogen bonding on AFP-ice interaction has diminished, previous models adopting the hydrophilic face of AFP as ice-binding face should be questioned. In an early study, Wen and Laursen[38] reported that mutating the Ala at the 5th position of the 11-residue repeats of wflAFP-6 to Gln or Leu resulted in complete loss of antifreeze activity. They proposed a model suggesting that the hydrophobic face was responsible for the dimer formation when wflAFP-6 accumulated onto ice surface. This model was not widely accepted as there was evidence showing that AFPs bind to the ice surface independently,[42] and recent studies using STM have demonstrated that type I AFPs bind individually to the ice surface.[43] Recently, Harding *et al.* (1999)[13] re-evaluated the data and proposed that the hydrophobic face is directly involved in interaction with the ice surface. This hypothesis was further supported by Baardsnes *et al.* (1999)[44] who demonstrated that mutating A17 or A21, which reside on the hydrophobic face, resulted in complete or significant loss of antifreeze activity, while mutating A19, which faces the hydrophilic side, had little or no effect. Nevertheless, the role of the hydrophilic face still cannot be completely ruled out. As mentioned above, introducing "-LT--N-" into wfsAFP enhances the activity by ~15%. Loewen *et al.* (1999)[45] have proposed that the Leu and Asn residues serve primarily to prevent aggregation and enhance AFP solubility and are not directly involved in ice interactions. However, the Asn→Gln mutation effectively destroyed antifreeze activity, demonstrating a possible steric restriction on the hydrophilic face. Thus, type I AFP-ice interaction may include residues from both the hydrophobic and hydrophilic faces clustered around the key Thr residues.

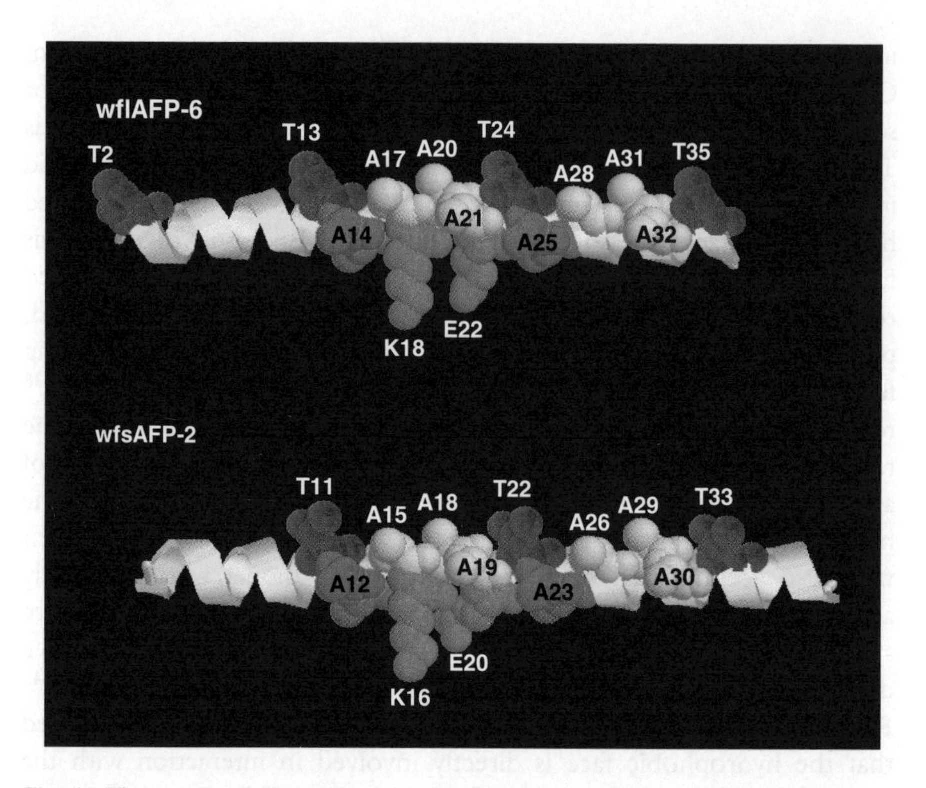

Fig. 6 The conserved 11-residue motif (TaaxAxxAAxx) residues form a similar face in the wflAFPs and wfsAFPs. Shown are the structure of wflAFP-6 as determined by X-ray crystallography,[33,59] and the structure of wflAFP-2 modeled as an ideal helix. The helix backbone is displayed as a ribbon, with the conserved residues (see Fig. 1 and text) of the motif shown in space-filling mode. The KE salt bridge in both AFPs is also shown for orientation purposes.

Potential Biological Roles of Skin-Type AFPs

The discovery of the skin-type AFPs has lead to some questions concerning their specific role in freezing protection. The primary role of the serum AF(G)Ps is believed to be non-colligative freezing point depression in the fish fluids by prevention of ice crystal growth. However, the possibility of intracellular AFPs in flounder skin is inconsistent with this role. There have been no reports on the physiological function of intracellular antifreezes, and the possibility of secretion of the skin-type AFPs outside the cell by some alternate secretion pathway has not yet been ruled out. However, antifreeze proteins and glycoproteins may play a wide variety of roles in freezing protection and cold tolerance. It is interesting to consider the possibility that some of these "secondary roles" may in fact be the primary functions of skin-type AFPs.

One possible role for antifreeze proteins would be to mask or inhibit the activity of endogenous or exogenous heterogenous ice nucleators. The ability of fish antifreezes and insect thermal hysteresis proteins to reduce the efficiency of heterogenous ice nucleation sites has been demonstrated.[46] Furthermore, an investigation of beetle larvae (*Dendroides canadensis*) indicates that fish antifreeze proteins can inhibit endogenous ice nucleators isolated from *D. canadensis* hemolymph.[47] It has also been suggested that the plasma membrane is capable of nucleating ice formation.[48] Thus, if the skin AFPs are intracellular, then they might play a role in preventing intracellular ice nucleation.

Even in freeze-tolerant animals, it is believed that ice growth must be restricted to the extracellular spaces,[49] although there are a minority of examples of intracellular ice formation. Extensive ice formation is unlikely in teleost fish that have antifreeze proteins and are freeze-resistant. But, small and localized freezing events cannot be ruled out. One mechanism of freezing damage at the cellular level is thought to be damage of the plasma membrane due to osmotic stresses caused by the growth of extracellular ice.[50,51] As the extracellular ice grows, a gradient of osmotic pressure is created across the membrane as solutes are excluded from the growing ice crystal. At a critical gradient, the membrane is damaged and intracellular freezing occurs, as the membrane no longer functions as a barrier to ice.[50] Furthermore, it has been demonstrated that cell-cell contacts, such as those found in intact tissues, affect the integrity of membranes and the formation of intracellular ice.[52,53] In these studies, when cells with various forms of intercellular contacts were assayed for intracellular ice formation, the probability of intracellular ice formation increased with increasing cell contacts and it was concluded that ice nucleation occurs through junctions between cells.[53] As well, in a study by Berger and Uhrík (1996),[54] intracellular ice was observed to propagate from cell to cell in cell strands from salivary glands. The authors concluded that the intracellular ice was progressing by way of the intercellular contacts between the adjacent cells.[54] Thus, one key factor in preventing freezing damage at the cellular level is to prevent the propagation of extracellular ice across the plasma membrane. If the cell membrane is damaged, it not only allows ice nucleators into the cell, it may also compromise the adjoining cells of the particular tissue through its intercellular contacts.

Considering the susceptibility of the cell membrane to freezing damage, it is interesting to note that the antifreeze proteins may have a role in maintaining the membrane barrier itself. Antifreeze proteins have been demonstrated to be able to prevent leakage of liposomes as they are cooled through the phase transition temperature[55] (and reviewed elsewhere in this volume). The extracellular and intracellular antifreeze proteins may act in conjunction to prevent membrane leakage and maintain membrane stability at low temperatures. The extracellular antifreezes would also prevent the osmotic stresses on plasma membranes by preventing ice crystal growth, thus maintaining membrane stability. The intracellular antifreeze proteins may also be responsible for maintaining the stability of the various organellar membranes. If a transient local breach of the plasma membrane were to occur, that would allow heterogenous ice nucleators access to the cellular compartment, their nucleating capabilities could be quickly masked by the presence of skin-type AFPs, thus minimizing cellular damage. Furthermore, skin-type AFPs may be necessary to prevent the propagation of ice from a damaged cell into an undamaged cell through intercellular contacts that would be found in a tissue system.

It is conceivable that the intracellular skin-type AFPs act in conjunction with the circulating extracellular AFPs to prevent membrane damage at peripheral tissues. This would be advantageous at the peripheral tissues such as the skin and gills, as these tissues are most likely the sites that would encounter the most exogenous ice nucleators and would have to act as a barrier to environmental ice. In fact, fish skin has been demonstrated to be an effectual barrier to ice propagation, a property that is enhanced in the presence of antifreeze proteins.[56] Thus, membrane protection and maintenance of membrane stability may be a role for intracellular antifreeze proteins.

Common Themes of Skin-Type AFPs

The identification of skin-type AFPs in three different fish species makes it possible to propose novel themes relating to this new subset of type I AFPs. First, the clones that have been isolated from the winter flounder, shorthorn sculpin and longhorn sculpin all lack signal peptide sequences and prosequences. Based on these features, it has been proposed that the skin-type AFPs may be intracellular. Second, the skin-type AFPs are found in a wide variety of tissues, with the highest levels of production in the peripheral tissues such as skin, gill filament and scales (winter flounder). These tissues would be in contact with external ice, and would require freezing protection. The third is that all skin-type AFPs fall into the larger biochemical class of type I AFPs.

The above themes are still general and largely untested. It will be interesting to re-visit them when a broader sampling of fish species that produce other classes of serum AFPs is performed. It may be that different species of fish have developed different adaptations from the three species discussed here. However, it appears that the skin-type AFPs are an important component of the freezing resistance in fish. Whether skin-type AFPs are a universal adaptation of all cold ocean fish species remains to be investigated.

Acknowledgements

We wish to thank Cora JL Young for assistance with preparation of the manuscript. Our work has been supported by the Medical Research Council (Canada) to CL Hew, the Natural Sciences and Engineering Research Council (Canada) to GL Fletcher and W Low is a recipient of an Ontario Graduate Scholarship.

References

1. O'Grady SM, Ellory JC and DeVries AL (1982). Protein and glycoprotein antifreezes in the intestinal fluid of polar fishes. *J. Expt. Biol.* **98**: 429–438.
2. Ahlgren JA, Cheng CC, Schrag JD and DeVries AL (1988). Freezing avoidance and the distribution of antifreeze glycopeptides in body fluids and tissues of Antarctic fish. *J. Expt. Biol.* **137**: 549–563.
3. Schneppenheim R and Theede H (1982). Freezing-point depressing peptides and glycoproteins from Arctic-Boreal and Antarctic fish. *Polar Biol.* **1**: 11–123.
4. Valerio PF, Kao MH and Fletcher GL (1990). Thermal hysteresis activity in the skin of the cunner, Tautogolabrus adspersus. *Can. J. Zool.* **68**: 1065–1067.
5. Gong Z, Fletcher GL and Hew CL (1992). Tissue distribution of fish antifreeze protein mRNAs. *Can. J. Zool.* **70**: 810–814.
6. Gong Z, Ewart KV, Hu Z, Fletcher GL and Hew CL (1996). Skin antifreeze protein genes of the winter flounder, Pleuronectes americanus, encode distinct and active polypeptides without the secretory signal and prosequences. *J. Biol. Chem.* **271**: 4106–4112.
7. Low WK, Miao M, Ewart KV, Yang DSC, Fletcher GL and Hew CL (1998). Skin-type antifreeze protein from the shorthorn sculpin, *Myoxocephalus scorpius*. *J. Biol. Chem.* **273**: 23098–23103.
8. Pickett M, Scott G, Davies P, Wang N, Joshi S and Hew C (1984). Sequence of an antifreeze protein precursor. *Eur. J. Biochem.* **143**: 35–38.
9. Davies PL, Roach AH and Hew CL (1982). DNA sequence coding for an antifreeze protein precursor from winter flounder. *Proc. Natl. Acad. Sci. USA* **79**: 335–339.
10. Hew CL and Yang DS (1992). Protein interaction with ice. *Eur. J. Biochem.* **203**: 33–42.
11. Davies PL and Hew CL (1990). Biochemistry of fish antifreeze proteins. *FASEB J.* **4**: 2460–2468.
12. Davies PL and Sykes BD (1997). Antifreeze proteins. *Curr. Opin. Struct. Biol.* 7: 828–834.
13. Harding MM, Ward LG and Haymet AD (1999). Type I "antifreeze" proteins. Structure-activity studies and mechanisms of ice growth inhibition. *Eur. J. Biochem.* **264**: 653–665.

14. Gong Z, King MJ, Fletcher GL and Hew CL (1995). The antifreeze protein genes of the winter flounder, *Pleuronectes americanus*, are differentially regulated in liver and non-liver tissues. *Biochem. Biophys. Res. Comm.* **206**: 387–392.
15. Davies PL and Gauthier SY (1992). Antifreeze protein pseudogenes. *Gene* **112**: 171–178.
16. Hew CL, Joshi S, Wang NC, Kao MH and Ananthanarayanan VS (1985). Structures of shorthorn sculpin antifreeze polypeptides. *Eur. J. Biochem.* **151**: 167–172.
17. Hew CL, Fletcher GL and Ananthanarayanan VS (1980). Antifreeze proteins from the shorthorn sculpin, *Myoxocephalus scorpius*: isolation and characterization. *Can. J. Biochem.* **58**: 377–383.
18. Fletcher GL, Addison RF, Hew CL and Slaughter D (1982). Antifreeze proteins in the Arctic shorthorn sculpin (*Myoxocephalus scorpius*). *Arctic* **35**: 302–306.
19. Cheng CH (1998). Evolution of the diverse antifreeze proteins. *Curr. Opin. Genet. Dev.* **8**: 715–720.
20. Deng G and Laursen RA (1998). Isolation and characterization of an antifreeze protein from the longhorn sculpin, *Myoxocephalus octodecimspinosis. Biochim. Biophys. Acta* **1388**: 305–314.
21. Deng G, Andrews DW and Laursen RA (1997). Amino acid sequence of a new type of antifreeze protein, from the longhorn sculpin *Myoxocephalus octodecimspinosis. FEBS Lett.* **402**: 17–20.
22. Zhao Z, Deng G, Lui Q and Laursen RA (1998). Cloning and sequencing of cDNA encoding the LS-12 antifreeze protein in the longhorn sculpin, *Myoxocephalus octodecimspinosis. Biochim. Biophys. Acta* **1382**: 177–180.
23. Low WK, Lin Q, Stathakis C, Miao M, Fletcher GL and Hew CL (2001). Isolation and characterization of skin-type, type I antifreeze polypeptides from the longhorn sculpin, *M. octodecemspinosus. J. Biol. Chem.* **276**(15): 11582–11589.
24. Powell JR and Moriyama EN (1997). Evolution of codon usage bias in Drosophila. *Proc. Natl. Acad. Sci. USA* **94**: 7784–7790.
25. Scott GK, Hew CL and Davies PL (1985). Antifreeze protein genes are tandemly linked and clustered in the genome of the winter flounder. *Proc. Natl. Acad. Sci. USA* **82**: 2613–2617.
26. Scott GK, Davies PL, Kao MH and Fletcher GL (1988). Differential amplification of antifreeze protein genes in the pleuronectinae. *J. Mol. Evol.* **27**: 29–35.

27. Devries AL and Lin Y (1977). Structure of a peptide antifreeze and mechanism of adsorption to ice. *Biochim. Biophys. Acta.* **495**: 388–392.
28. Knight CA, Cheng CC and DeVries AL (1991). Adsorption of alpha-helical antifreeze peptides on specific ice crystal surface planes. *Biophys. J.* **59**: 409–418.
29. Chao H, Houston ME, Jr., Hodges RS, Kay CM, Sykes BD, Loewen MC, Davies PL and Sönnichsen FD (1997). A diminished role for hydrogen bonds in antifreeze protein binding to ice. *Biochemistry* **36**: 14652–14660.
30. Haymet AD, Ward LG, Harding MM and Knight CA (1998). Valine substituted winter flounder "antifreeze": preservation of ice growth hysteresis. *FEBS Lett.* **430**: 301–306.
31. Haymet AD, Ward LG and Harding MM (1999). Winter flounder "antifreeze" proteins: synthesis and ice growth inhibition of analogues that probe the relative importance of hydrophobic and hydrogen-bonding interactions. *J. Am. Chem. Soc.* **121**: 941–948.
32. Wierzbicki A, Taylor MS, Knight CA, Madura JD, Harrington JP and Sikes CS (1996). Analysis of shorthorn sculpin antifreeze protein stereospecific binding to (2–1 0) faces of ice. *Biophys. J.* **71**: 8–18.
33. Lin Q, Ewart KV, Yan Q, Wong WK, Yang DS and Hew CL (1999). Secretory expression and site-directed mutagenesis studies of the winter flounder skin-type antifreeze polypeptides. *Eur. J. Biochem.* **264**: 49–54.
34. Sicheri F and Yang DSC (1995). Ice-binding structure and mechanism of an antifreeze protein from winter flounder. *Nature* **375**: 427–431.
35. Lin Q, Ewart KV, Yang DS and Hew CL (1999). Studies of a putative ice-binding motif in winter flounder skin-type antifreeze polypeptide. *FEBS Lett.* **453**: 331–334.
36. Tong L, Lin Q, Wong WK, Ali A, Lim D, Sung WL, Hew CL and Yang DS (2000). Extracellular expression, purification, and characterization of a winter flounder antifreeze polypeptide from *Escherichia coli. Protein Exprt. Purif.* **18**: 175–181.
37. Chakrabartty A and Hew CL (1991). The effect of enhanced alpha-helicity on the activity of a winter flounder antifreeze polypeptide. *Eur. J. Biochem.* **202**: 105–1063.
38. Wen D and Laursen RA (1993). Structure-function relationships in an antifreeze polypeptide. The effect of added bulky groups on activity. *J. Biol. Chem.* **268**: 16401–16405.

39. Houston ME, Jr., Chao H, Hodges RS, Sykes BD, Kay CM, Sönnichesen FD, Loewen MC and Davies PL (1998). Binding of an oligopeptide to a specific plane of ice. *J. Biol. Chem.* **273**: 11714–11718.
40. Raymond JA and DeVries AL (1977). Adsorption inhibition as a mechanism of freezing resistance in polar fishes. *Proc. Natl. Acad. Sci. USA* **74**: 2589–2593.
41. Zhang W and Laursen RA (1998). Structure-function relationships in a type I antifreeze polypeptide. The role of threonine methyl and hydroxyl groups in antifreeze activity. *J. Biol. Chem.* **273**: 34806–34812.
42. Chao H, DeLuca CI and Davies PL (1995). Mixing antifreeze protein types changes ice crystal morphology without affecting antifreeze activity. *FEBS Lett.* **357**: 18–186.
43. Grandum S, Yabe A, Nakagomi K, Makoto T, Takemura F, Kobayashi Y, and Frivik PE (1999). Analysis of ice crystal growth for a crystal surface containing adsorbed antifreeze proteins. *J. Cryst. Growth* **205**: 382–390.
44. Baardsnes J, Kondejewski LH, Hodges RS, Chao H, Kay C and Davies PL (1999). New ice-binding face for type I antifreeze protein. *FEBS Lett.* **463**: 87–91.
45. Loewen MC, Chao H, Houston ME, Jr., Baardsnes J, Hodges RS, Kay CM, Sykes BD, Sönnichsen FD and Davies PL (1999). Alternative roles for putative ice-binding residues in type I antifreeze protein. *Biochemistry* **38**: 4743–4749.
46. Wilson PW and Leader JP (1995). Stabilization of supercooled fluids by thermal hysteresis proteins. *Biophys. J.* **68**: 209–2107.
47. Olsen TM and Duman JG (1997). Maintenance of the supercooled state in overwintering pyrochroid beetle larvae, Dendroides canadensis: Role of hemolymph ice nucleators and antifreeze proteins. *J. Comp. Physiol. B Biochem. Syst. Environ. Physiol.* **167**: 105–113.
48. Toner M and Cravalho EG (1990). Thermodynamics and kinetics of intracellular ice formation during freezing of biological cells. *J. Appl. Phys.* **67**: 1582–1593.
49. Lee REJ and Costanzo JP (1998). Biological ice nucleation and ice distribution in cold-hardy ectothermic animals. *Ann. Rev. Physiol.* **60**: 55–72.
50. Muldrew K and McGann LE (1994). The osmotic rupture hypothesis of intracellular freezing injury. *Biophys. J.* **66**: 532–541.

51. Muldrew K and McGann LE (1990). Mechanisms of intracellular ice formation. *Biophys. J.* **57**: 525–532.
52. Acker JP and McGann LE (2000). Cell-cell contact affects membrane integrity after intracellular freezing. *Cryobiology* **40**: 54–63.
53. Acker JP, Larese A, Yang H, Petrenko A and McGann LE (1999). Intracellular ice formation is affected by cell interactions. *Cryobiology* **38**: 363–371.
54. Berger WK and Uhrik B (1996). Freeze-induced shrinkage of individual cells and cell-to-cell propagation of intracellular ice in cell chains from salivary glands. *Experientia* **52**: 843–850.
55. Hays LM, Feeney RE, Crowe LM, Crowe JH and Oliver AE (1996). Antifreeze glycoproteins inhibit leakage from liposomes during thermotropic phase transitions. *Proc. Natl. Acad. Sci. USA* **93**: 6835–6840.
56. Valerio PF, Kao MH and Fletcher GL (1992). Fish skin: an effective barrier to ice crystal propagation. *J. Exp. Biol.* **164**: 135–151.
57. Page RD (1996). TreeView: an application to display phylogenetic trees on personal computers. *Comput. Appl. Biosci.* **12**: 357–358.
58. Thompson JD, Gibson TJ, Plewniak F, Jeanmougin F and Higgins DG (1997). The CLUSTAL_X windows interface: flexible strategies for multiple sequence alignment aided by quality analysis tools. *Nucleic Acids Res.* **25**: 4876–4882.
59. Jeanmougin F, Thompson JD, Gouy M, Higgins DG and Gibson TJ (1998). Multiple sequence alignment with Clustal X. *Trends Biochem. Sci.* **23**: 403–405.
60. Yang DS, Sax M, Chakrabartty A and Hew CL (1988). Crystal structure of an antifreeze polypeptide and its mechanistic implications. *Nature* **333**: 232–237.

Chapter 8

The Interaction of Antifreeze Proteins with Model Membranes and Cells

Melanie M Tomczak and John H Crowe
Section of Molecular and Cellular Biology
University of California
One Shields Ave
Davis, CA 95616, USA

Introduction

Many cells are damaged when they are chilled below their physiological temperatures. For instance, temperate zone plants such as tomatoes wilt when exposed to temperatures below about 10°C.[1] Similarly, human blood platelets spontaneously activate when they are chilled below room temperature.[2,3] Chilling damage of this sort has widely been ascribed to lipid phase transitions in cell membranes,[4–9] although this suggestion has been controversial.[10] Nevertheless, it is well known that lipid membranes become transiently leaky as they pass through lipid phase transitions[11–13] and that the phase transitions can lead to lateral phase separations of membrane components,[14] including aggregation of membrane proteins.[15] More recently, chilling has been shown to lead to formation of discreet membrane domains known as "rafts".[16]

This is a matter of considerable practical significance. Temperate zone plants are restricted in their growing season and geographic distribution in no small part because of chilling damage. Human blood platelets are stored at room temperature because they are extremely cold-sensitive. Room temperature storage only allows the platelets to be kept for a short time. Because they are discarded after a maximum of five days, there is a chronic shortage of platelets in blood banks. Thus, efforts have been underway for decades to discover means for protecting cells against chilling damage.

The first suggestion in the literature that antifreeze proteins and glycoproteins [AF(G)Ps] might be involved in inhibition of chilling damage came from work by Amir Arav, then a postdoctoral researcher with Boris Rubinsky at the University of California, Berkeley. They serendipitously discovered that AFGPs inhibit damage to pig oocytes during chilling to hypothermic temperatures.[17] They suggested that the AFGPs block ion channels and thereby prevent ion leakage across cell membranes when they are chilled to low temperatures. Subsequently, Hays and colleagues showed that AFGPs prevent leakage across model membranes as they are chilled through their thermotropic phase transition.[18] Since these membranes were composed of phospholipids alone, it became clear that the effect first ascribed by Rubinksy *et al.* (1990)[17] to an interaction between the AFGPs and ion channels is more likely due to an interaction between the peptides and the lipid bilayer. In this review, we will discuss the proposed hypotheses for the mechanism of protection of membranes by antifreeze proteins during chilling, summarize our current state of knowledge and speculate where the field of antifreeze protein-mediated membrane stabilization may be headed.

Membrane Phase Transitions and Low Temperature Damage

Membrane lipids are usually in the fluid, liquid crystalline phase at physiological temperatures, which is characterized by free rotational and lateral mobility of the lipid molecules within the bilayer. The lipid spacing allows hydrogen bond interactions between the headgroups and surrounding water, and allows the acyl chains to be in a *gauche* conformation with relatively high mobility [reviewed in Lewis and McElhaney (1991)].[19] Lipids undergo a phase transition from the liquid crystalline to the more rigid gel phase as membranes are cooled. The gel phase is characterized by tighter packing of the lipid headgroups and the acyl chains. In this phase, the acyl chains become extended and take on an all-*trans* conformation [reviewed in Gennis (1989)],[20] which is maintained by van der Waals and steric interactions.[19] The closer

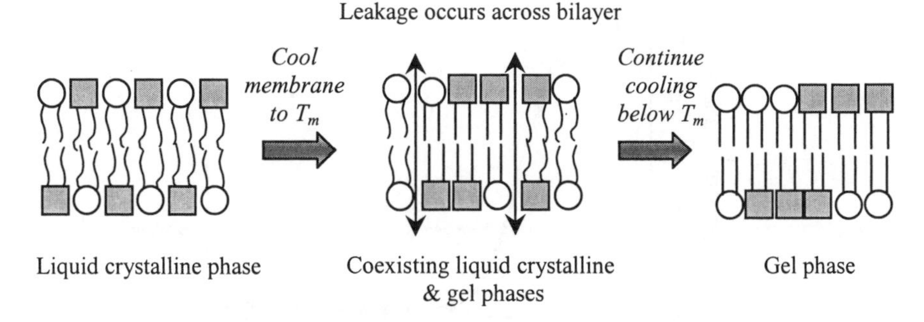

Fig. 1 Leakage occurs across bilayers during the phase transition. A membrane (left) is drawn in the fluid, liquid crystalline phase. As the membrane is cooled, there is transient leakage across the bilayer during the phase transition (middle). Below the phase transition, the membrane is in the more rigid gel phase (right).

headgroup packing excludes water from between the lipid molecules and causes interactions between adjacent headgroups.

Since the phase transition is not completely cooperative, even in bilayers composed of a single species of pure phospholipids, liquid crystalline and gel phase domains co-exist transiently during chilling. The phase boundaries between gel and liquid crystalline domains are thought to be the site of packing defects, leading to leakage across the bilayer (Fig. 1).[11–13] Clearly, such leakage is damaging to cells and tissues because ion gradients would be dissipated, for example. Chilling-induced leakage can be observed easily in model systems employing liposomes that contain a self-quenching fluorescent dye, such as carboxyfluorescein (CF). The dye, which is trapped at high concentration in the aqueous interior of the liposomes, leaks out of the liposomes as they are cooled through their phase transition, and into the bulk solution, where it is diluted and thus fluoresces. The resulting fluorescence can be measured as an indication of damage to the membranes (Fig. 2).[18] This leakage is transient, and is abated after the transition is complete. There is also a strong kinetic component to the leakage; the longer the membrane is in the phase transition, the greater the leakage (Hays *et al.*, in preparation).

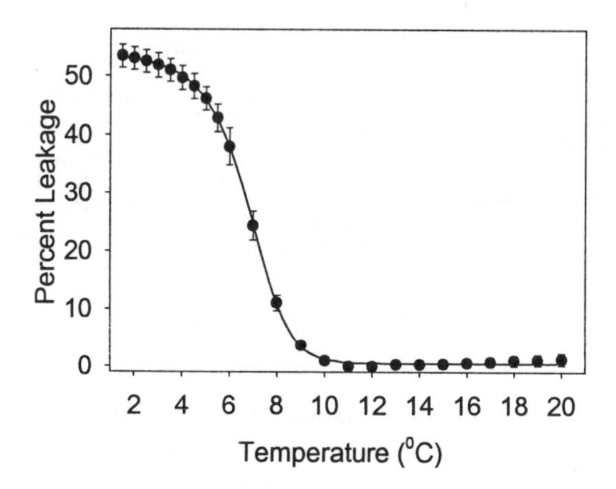

Fig. 2 Leakage from liposomes as they are cooled through their phase transition. DEPC liposomes were cooled through their T_m (12°C). Carboxyfluorescein, a self-quenching fluorescent dye trapped in the aqueous interior of the liposomes, leaked out of the liposomes during the phase transition. The resulting fluorescence was measured and percent leakage was determined as a function of temperature according to Hays *et al.* (1996). Data courtesy of LM Hays.

Determinants of Lipid Phase Transitions

Not all cells are chilling sensitive, and the reasons for this are reasonably well understood; lipid alterations can affect the phase transition temperature (reviewed in Refs. 10 and 21). Long chain lipids have a higher phase transition temperature (T_m) because there are more van der Waals interactions between the acyl chains to keep them aligned and, accordingly, a larger amount of heat is required for a phase change to the liquid crystalline state to occur.[19] The opposite is true for short chain lipids, which tend to have low T_m.

Another determinant of T_m is the degree and location of unsaturation in the acyl chains. A highly unsaturated lipid will have a lower phase transition temperature due to disorder among the acyl chains induced by the double bond kinks. However, double bonds have less of an effect on T_m when they are located at the top or bottom of the acyl chains, or if they are trans double bonds because these situations allow relatively more order in acyl chain packing.[19]

Headgroup charge plays a role in determining the T_m of a lipid. Lipids with charged headgroups, for example phosphatidylserine, have greater spacing between adjacent lipid molecules due to like-charge repulsion. This headgroup repulsion decreases the T_m of the lipids.[22] If counter ions, such as Ca^{2+}, are present in the solution the headgroups are pulled closer together due to charge balance, and the T_m is increased.

Cholesterol is a great disturber of phase transitions. It increases fluidity below T_m because it inhibits the all-*trans* conformation, and decreases fluidity above T_m by decreasing the fraction of *gauche* conformers.[20] This results in a broader phase transition and, if the cholesterol concentration is high enough, it will entirely mask the transition.[23] Recently, cellular membrane domains, or "rafts," that are rich in cholesterol, sphingomyelin and certain signal transducing integral membrane proteins have been characterized.[16] These domains grow in size as the temperature decreases (Gousett *et al.*, submitted), probably due to lateral phase separation of like membrane components. Low temperature lateral phase separation is thought to be a major cause of

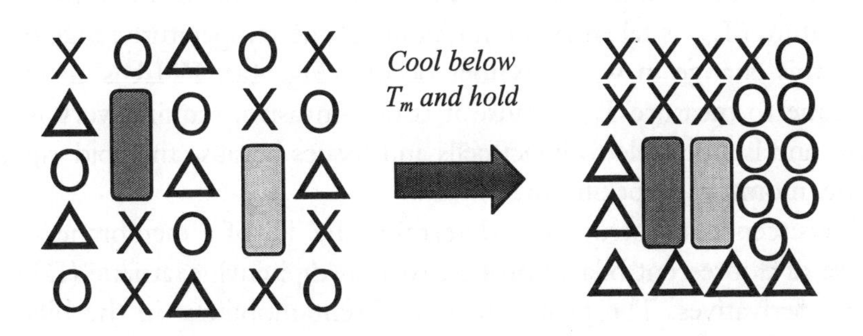

Fig. 3 Lateral phase separation in a membrane after cooling. This diagram represents a membrane composed of three types of lipid (X, O, Δ) and two integral membrane proteins (rectangles) at a physiological temperature (left). As the membrane is cooled, its components de-mix into domains of like molecules (right). During rewarming, the membrane components will not necessarily remix, rendering the cell non-functional.

damage in chilling sensitive cells[14] because integral membrane proteins tend to aggregate and the domains of like lipid components do not completely remix in the membrane upon rewarming (Fig. 3), rendering the cells non-functional.[15] Recent data suggest that rafts play a role in human platelet activation, the first evidence that rafts have physiological functions in cells (K. Gousset *et al.*, in preparation).

Methods for Determining Membrane Phase Transitions

Fourier transform infra-red (FTIR) spectroscopy measures the vibration of chemical groups in a sample, each of which has a characteristic wave number under specific conditions. The symmetric CH_2 stretch of the fatty acyl chains is monitored as a function of temperature to determine the T_m of a membrane with FTIR. A CH_2 stretch wave number characteristic of the liquid crystalline phase is near 2583 cm^{-1}, and decreases with temperature until the membrane is in the gel phase, where the wave number is near 2850 cm^{-1}.[24] The phase transition temperature (T_m) is taken as the midpoint of the temperature range over which this change in wave number occurs (Fig. 4a). FTIR is an ideal technique to measure T_m because it is non-invasive, requires very little sample and is amenable to intact cells and tissues because the lipid signal can be monitored specifically.[5,24]

Fluorescence anisotropy can determine the T_m of a membrane with the use of fluorescent bilayer probes, such as diphenylhexatriene (DPH) and its derivatives. The probes have different mobilities in the bilayer based on the phase of the membrane. Polarized light is used to excite the probes oriented in the direction parallel to the light and the probe anisotropy can be calculated.[25] A probe has greater mobility when the membrane is in the liquid crystalline phase and, therefore, a lower anisotropy value. Accordingly, the probe is more rigidly held in place in the gel phase and the resulting anisotropy value is higher (Fig. 4b). As with FTIR, the T_m is taken as the midpoint of the temperature range over which the change in anisotropy occurs. Two drawbacks

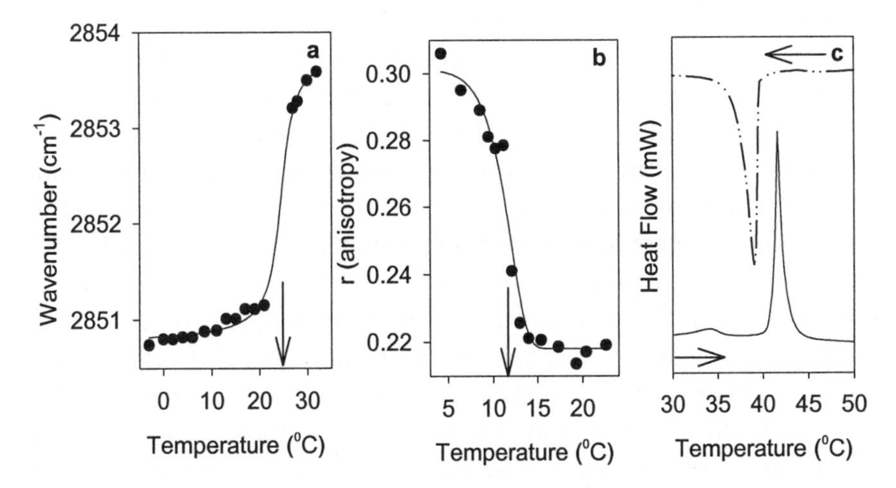

Fig. 4 Methods for determining membrane phase transitions. **(a)** Fourier transform infrared spectroscopy (FTIR). The frequency of the symmetric CH_2 stretch of the dimyristoylphosphatidylcholine (DMPC) acyl chains is plotted as a function of temperature. The midpoint of the change in wave number over temperature is taken as the phase transition temperature, T_m (arrow). **(b)** Fluorescence anisotropy. The anisotropy (r) of trimethylammonium-diphenylhexatriene (TMA-DPH) is plotted as a function of temperature for DEPC liposomes during warming. Again, the T_m is taken as the midpoint of the change in anisotropy over temperature (arrow). **(c)** Differential scanning calorimetry (DSC). The differential heat flow between the DPPC sample and an inert reference is plotted as a function of temperature as the two are heated and cooled simultaneously. The T_m is taken as the peak of the heat flow plot. DSC data courtesy of LM Crowe.

of fluorescence anisotropy, in comparison with FTIR, are that a larger sample is needed and that a membrane probe must be incorporated into the bilayer.

The last method for measuring T_m that will be discussed here is differential scanning calorimetry (DSC). DSC compares the enthalpy change of a membrane sample in relation to an inert reference as the two are heated simultaneously.[19] The differential heat flow between the two samples is plotted as a function of temperature. For example, melting of a lipid from the gel to liquid crystalline phase is an endothermic event, and the lipid sample will require more heat than the reference to maintain the same temperature (Fig. 4c). An exothermic event is observed when a lipid is cooled from the liquid crystalline to the gel phase, as more order is achieved and heat is released. The T_m is taken as the peak of the heat flow plot. There is a hysteresis between the cooling and warming T_m because the lipids require slightly more heat to melt from an ordered state and, so, the resulting warming T_m is at a slightly increased temperature compared with the cooling T_m. DSC is a good technique for studying model membranes that contain only one or a few components that have cooperative transitions. However, it is not as amenable to the measurement of cellular phase transitions because cell membranes are composed of many types of lipids as well as integral membrane proteins and, as more components are added to a system, the transition broadens and a clear T_m is not necessarily observed.

Antifreeze Proteins from Cold Hardy Non-Fish Species: Plants and Overwintering Insects

This review will focus primarily on membrane stabilization by fish antifreeze proteins, but first we wish to discuss briefly antifreeze proteins (AFPs) from cold hardy plants and overwintering insects. We are unaware of any investigations into whether AFPs from cold hardy plants might stabilize membranes under low temperature stress. However, this is a distinct possibility. Carrots, winter wheat and rye, bittersweet nightshade and other cold hardy plants produce AFPs that are no more active than the fish AFPs.[26–30] These plants survive very low winter temperatures beyond the maximal freezing point depression of these proteins, which depress the freezing point of solution by only a fraction of a degree.[26–30] Thus, it is difficult to see how the proteins would have much use as antifreezes. Additionally, plants that overwinter are freeze-tolerant, not freeze-avoiding, again suggesting that preventing ice crystal growth is not the role for these proteins in the plant. The plant AFPs are good ice recrystallization inhibitors, and may play a role in preventing the ice crystal growth-induced damage that would normally occur on rewarming to temperatures below freezing. It is also possible that these proteins prevent damage caused to membranes during low temperature and freezing stress.

There is only one published study on cellular stabilization by an insect AFP. Until recently, the AFPs of insects were called thermal hysteresis proteins (THPs). A THP from *Dendroides canadensis* was added to gut cells of a freeze-tolerant centipede, *Lithobius forficatus*, and survival after freezing was determined.[31] This centipede makes a THP similar to the one from *D. canadensis* utilized in this study, so control centipedes collected in the summer and winter, where the protein would be naturally absent or present, were included in the study. In all cases, the *D. canadensis* THP decreased the LT_{50} for the cells, including the winter gut cells that already contained the native THP. Interestingly, when the THP was added to summer cells and they were washed to remove any protein not adhering, the LT_{50} decreased to almost the same level as the cells that were not washed, indicating that the protein adherence to the cells may be involved in providing protection. Cellular stabilization by THPs may be one part of the freezing protection, but the authors speculate that there must be additional factors that aid in freezing tolerance in these centipedes.[31]

Hypothermic Storage of Cells with Antifreeze Proteins

Pig oocytes were used in the first reported study of hypothermic membrane stabilization by an antifreeze protein[17] because they are unable to survive exposure to temperatures below 10°C.[32] The porcine oocytes were chilled to 4°C for four or 24 hours with increasing concentrations of AFGPs, and recovery was determined by monitoring the resulting membrane potential. After four hours, only 20% of the chilled control oocytes had a membrane potential within two standard deviations of the fresh control oocytes. After 24 hours, the number dropped to zero, confirming that the cells are damaged by low temperature. In contrast, oocytes chilled in the presence of 1 or 40 mg/ml of AFGP fractions 1–5 + 7–8 had roughly 80% and 60% survival after four and 24 hours, respectively. Those chilled with

0.1 mg/ml fractions 1–5 + 7–8, or 40 mg/ml fractions 1–5 or fractions 7–8 had an approximately 40% survival rate after four hours, but none survived a 24-hour exposure.

These results led to the proposal of a multi-faceted mechanism for the stabilization of oocytes by AF(G)Ps. First, they reported that the oolemma remained intact after hypothermic storage with AFGPs and suggested that the peptides protect the structure of the oolemma and inhibit leakage from the oocyte during chilling.[17] They continued by proposing that the AFGPs block ion channels under hypothermic conditions by binding to the hydrophilic portions of membrane proteins, presumably through the sugar moieties on the peptides. The authors state that the effect was typical of a protein-protein interaction because it was nonlinear: 1 mg/ml and 40 mg/ml peptide solutions offered the same amount of protection to the oocytes. However, the authors only reported results from three protein concentrations: 0.1 mg/ml, where there was no protection after 24 hours, and 1.0 and 40 mg/ml, where there was equal protection after 24 hours. With so few data points, it is impossible to determine if the effect was, indeed, nonlinear. These authors also did not consider the fact that the peptides could be protecting the oocytes from the damage normally induced by lipid phase transitions. One important point is that the authors stated neither the volumes of their samples nor the concentration of oocytes in each sample.[17] This, along with more protein concentrations, would have been helpful, in retrospect, for determining the optimal ratio of protein:oocyte needed for protection.

The next question was whether other AF(G)Ps possess similar membrane-stabilizing properties. Three non-glycosylated fish antifreeze proteins (AFPs), termed AFP I, II and III, had been identified by 1991.[33] These proteins were tested on bovine oocytes to determine whether they offered protection to cells during storage at 4°C for 24 hours[34] in a manner similar to the AFGPs.[17] *In vitro* maturation occurred in 80% of the fresh oocyte controls and 55% were fertilized *in vitro*, whereas in the chilled controls, only 24% underwent maturation and 0% were fertilized. AFP type I, II and III increased *in vitro* maturation to roughly 70%, and 38–48% of the oocytes were fertilized after 24 hours at 4°C. These results implied that all known fish AF(G)Ps protected membranes as they are chilled to hypothermic, non-freezing temperatures. The authors again suggested that the AFPs conferred low temperature protection by blocking ion channels.[34]

The possibility of low temperature stabilization of an intact rat liver by an AFP was investigated. The liver was perfused with a solution of AFP type III and stored at low temperature.[35] Subsequently, physiological functions were monitored compared to the control. Two measures of liver function were used to determine recovery: (1) the secretion of the cytosolic enzyme lactate dehydrogenase (LDH) as an indicator of cell death; and (2) the production of bile over time as an indicator of a functioning liver. Production of bile after low temperature storage with AFP type III was nearly three times that of the liver stored without AFP, but it was only half that of the fresh liver. However, the secretion of LDH in the liver stored with AFP was similar to that of the fresh control, suggesting that the peptide protected membrane integrity in the liver during low temperature exposure. These results suggest that the peptide protected the cells, but that it did not stabilize the organ ultrastructure, so full functionality of the liver was not maintained.

Rubinsky and co-workers proposed that hypothermic membrane stabilization was a fundamental property of antifreeze proteins[34] because every AF(G)P tested appeared to protect membranes during low temperature exposure. They suggested that the AF(G)Ps might have initially evolved to protect fish membranes when the temperature of ocean was decreasing but before the fish needed protection against ice crystals. They proposed that the proteins subsequently evolved their current activity of ice growth inhibition as the polar waters became ice laden. This hypothesis is intriguing in view of the recent discovery of a number of intracellular antifreeze proteins homologous to AFP type I.[36] These AFPs are found inside cells during specific periods of development and have lower antifreeze activity (thermal hysteresis) than the serum type I AFPs. It is possible that these AFPs mediate *in vivo* low temperature-membrane stabilization in the fish because of these characteristics, although we are unaware of any investigation into this idea.

Toxicity of Antifreeze Proteins

There are several reports in the literature that AF(G)Ps are actually toxic, at least in certain specialized cells.

Human Oocytes

Another trial for the AFPs was low temperature stabilization of human oocytes.[37] Here, the oocytes were chilled in the presence of either 1 or 10 mg/ml AFP type I or type III for 20 hours at 4°C. With 1 mg/ml AFP type I or type III, 90% of the oocytes were fertilized. However, a toxic effect of the peptides at a concentration of 10 mg/ml was observed. Less than 30% of the oocytes were capable of being fertilized, the oocytes had an irregular shape and the ooplasm was granulated after incubation at the higher peptide concentration. The authors suggested that the damage at high concentration came from irreversible binding to ion channels so, after rewarming, the channels did not function properly and the cells died.

However, it could be reasoned that if there is 90% survival with a lower concentration of peptide and the mechanism of action is to block ion channels, then they should have all been blocked with the lower concentration and any additional peptide would have no place to bind. That was not the case in this study. We suggest that this toxicity may be due to AFP/AFGP–lipid interactions that cause disruptions in the bilayer structure, including the promotion of membrane fusion (MMT *et al.*, in preparation).

Plant Thylakoids

In another study, AF(G)Ps were found to be toxic to certain types of membranes under specific conditions.[38] These authors reported that AFGPs and AFP I and III are cryotoxic to thylakoid membranes, which house the photosynthetic machinery of the plant. In this study, freezing damage was measured as leakage of the soluble thylakoid lumen protein plastocyanin (PC). PC leaks from thylakoids in a time-dependent, biphasic manner after freezing: an initial rapid leakage (t = 1/2 h) of roughly 30% of the total plastocyanin, followed by a slow, steady leakage of ~10% more PC (t_{final} = 6 h). AFGPs and AFP type I and III increased the initial rate of leakage in a concentration-dependent manner. AFP type III (1.0 mg/ml) was the most damaging, increasing the initial leakage to 90%, and 1.0 mg/ml AFP type I or AFGPs caused the initial leakage to increase to nearly 70%. Only the AFGPs increased the slow phase of PC release, resulting in 85% total PC leakage after freezing for five hours.

These data indicate that, although all three AF(G)Ps are damaging to thylakoids after freezing, it is possible that they mediate the damage in different manners. The authors also report that at very low concentrations (5–50 μg/ml), the proteins offered no protection during freezing, in contrast to findings with vitrification of red blood cells.[39] In addition, they found that AFP type III and the AFGPs caused a small increase in PC leakage when stored at 0°C over time, suggesting that these proteins do not need freezing conditions to interact with the membranes.[38] The reader must be cautioned that the thylakoid results cannot be directly extrapolated to other systems because galactolipids make up 80% of the thylakoid lipids,[40] resulting in a membrane topology that is very different from that of animal cells.

Decreased Membrane Permeability at Room Temperature

Two reports suggest that AFP type I and type III interact with cell membranes to block ion channels in the absence of low temperature. In the first study, the membrane potential of pig granulosa cells was measured in the presence of AFP type I at 22°C.[41] AFP type I at concentrations greater than 0.5 mg/ml completely inhibited Ca^{2+} and K^+ currents after depolarization, although the K^+ current decayed more slowly. The authors propose that the peptide was blocking the Ca^{2+} and K^+ channels differentially and preventing the ions from re-entering the cell after depolarization.

The second study used rabbit gastric parietal cells to test whether AFP type III decreased the calcium ion permeability of the plasma membrane after a carbachol-induced calcium spike.[42] The authors state that the control Ca^{2+} spike is followed by a high plateau of cytosolic calcium when carbachol and extracellular Ca^{2+} are present. However, there was no high plateau when the cells were spiked in the presence of >1.0 mg/ml AFP, suggesting that the AFP blocked the influx of extracellular Ca^{2+}. The rate of intracellular calcium decrease in cells incubated with or without AFP was the same in the absence of extracellular Ca^{2+}, indicating that the peptide had no effect on intracellular Ca^{2+} release. More recently, Oliver *et al.* (1999)[43] showed that AFGPs had no effect on the increase in cytosolic Ca^{2+} in chilled human blood platelets, although the peptides prevent cold-induced physiological activation of platelets [Tablin *et al.* (1996); see below].[8] Taken together, these data suggest that antifreeze proteins affect certain properties of plasma membranes, especially membrane permeability, but that theydo not appear to affect intracellular signaling cascades.

Stabilization of Membranes During Phase Transitions

Hays *et al.* (1996)[18] found that AFGPs inhibit leakage from phospholipid liposomes as they are chilled through their thermotropic phase transition. Importantly, this study showed that AFGPs and AFPs protect model membranes composed only of lipids.[18] They reasoned that the peptides interact with the lipid portion of the membrane to prevent the leakage that would normally occur as membranes are chilled through their T_m (see Fig. 2), instead of blocking ion channels as proposed by Rubinsky and co-workers.[17,34,35] Specifically, leakage across membranes with a T_m = 12°C was inhibited during chilling to 2°C by AFGP fraction 8 at 10 mg/ml or AFGP fraction 2–6 at 1 mg/ml (Fig. 5). These proteins decreased the membrane permeability coefficient by the same amount during cooling and warming, and the change in permeability was the same regardless of the membrane T_m.[18] This suggests that low temperature is not necessarily required for the peptides

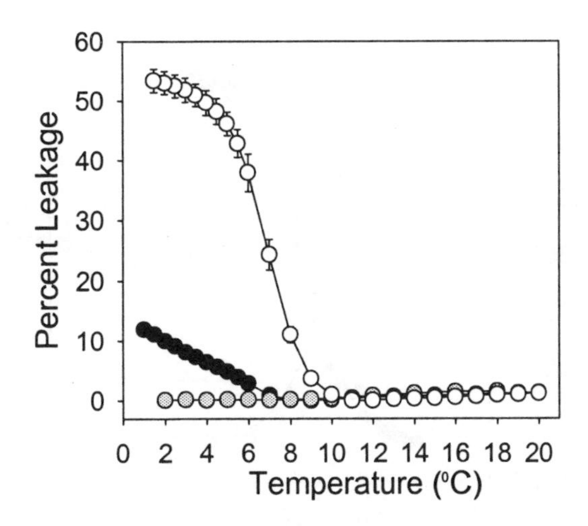

Fig. 5 Leakage is inhibited from phospholipid liposomes during chilling by AFGPs. Percent leakage from DEPC liposomes in the presence of 1 mg/ml AFGP fr. 2–6+ (black circles) or 10 mg/ml AFGP fr. 8 (gray circles) is plotted in comparison with the percent leakage from liposomes alone (open circles) as the liposomes are chilled through their T_m [adapted with permission from Hays *et al.* (1996)].[18]

to decrease membrane permeability but only a T_m or other change generating transient membrane defects.

A subsequent study supporting the hypothesis that antifreeze proteins are preventing damage normally incurred during T_m is the hypothermic protection of human blood platelets with AFGPs.[8] Physiological activation occurs in human platelets when they are chilled below 20°C. During cold-induced activation, platelets undergo a shape change from discoid to spherical and extend pseudopodia.[2,3] As well, the alpha granules inside the platelets fuse with the platelet plasma membrane and secrete their contents from the cell. These activation events are well correlated with the phase transition of intact platelets, indicating that the low temperature damage observed stems from the lipid phase change (Fig. 6).[8,9] The activation events are reversible after storage for 24 hours at 4°C, but are irreversible after 48 hours. AFGP fractions 5–7 prevent chilling-induced platelet activation after storage for three weeks at 4°C. Upon rewarming, the platelets were

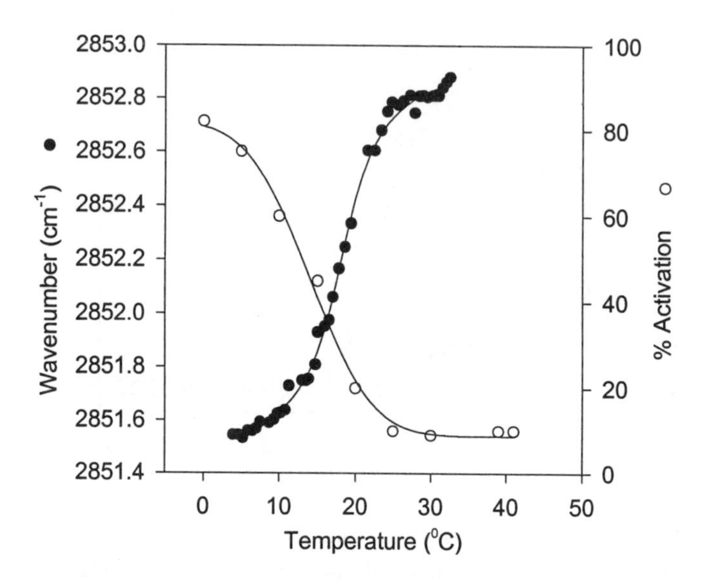

Fig. 6 Cold-induced platelet activation is correlated with membrane phase transitions [adapted with permission from Tablin *et al.* (1996)].[8]

morphologically normal and responded to agonists in a similar manner to fresh platelets.[8]

As mentioned above, cytosolic Ca^{2+} increases when platelets are chilled, however AFGPs have no effect on this step of activation.[43] This suggests that the AFGPs prevent low temperature damage to cells or liposomes through a topological or surface interaction, perhaps by holding the lipid molecules in place to prevent lateral phase separation and by blocking leakage at membrane defects. No change in phase transition temperature in the presence of AFGPs was detected, suggesting that an altered T_m is not the manner in which activation is prevented.[8]

Effects of AF(G)Ps on Membranes Containing Plant Galactolipids

Currently, we are investigating the effects of AFP type I on a model membrane system composed of equal weight fractions of

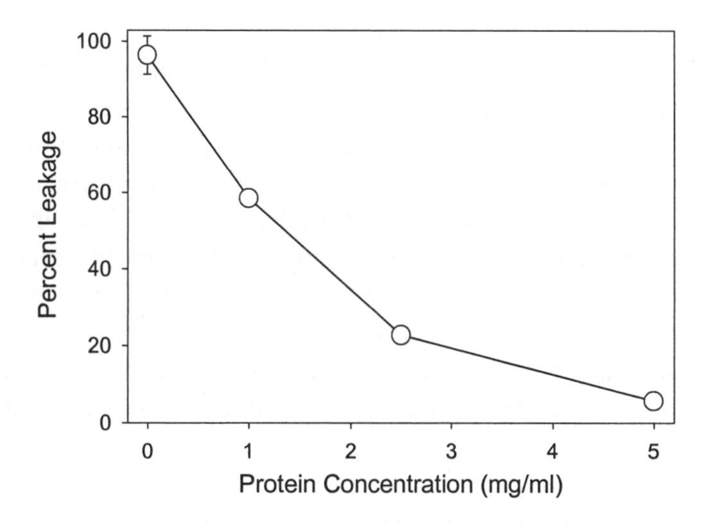

Fig. 7 Leakage is inhibited from galactolipid-containing liposomes in the presence of AFP type I. Percent leakage from DGDG:DMPC liposomes is plotted as a function of AFP type I concentration after the liposomes are cooled through their T_m.

dimyristoylphosphatidylcholine (DMPC) and the thylakoid lipid digalactosyldiacylglycerol (DGDG) during chilling (MMT *et al.*, submitted). AFP type I inhibits leakage from these liposomes in a concentration-dependent manner, with complete inhibition at 5 mg/ml (Fig. 7). AFGP fractions 6–8 or fraction 8 alone offered no protection to these liposomes. These data show that the peptides have a nearly opposite effect on galactolipid-containing liposomes as they have on pure phospholipid liposomes, where AFP type I only moderately inhibits leakage and the AFGPs completely protect the liposomes during chilling (Fig. 5).[18]

Mechanisms of Interaction

One hypothesis for the mechanism by which AF(G)Ps decrease permeability of membranes during chilling was that they interact directly with the membrane to lower the permeability. Previously, there was no evidence to show a direct interaction. Now we have several lines of evidence to support the hypothesis that AFP type I interacts directly with the membranes. First, AFP type I significantly increases the T_m of the DGDG:DMPC liposomes, as measured by FTIR and fluorescence anisotropy (Fig. 8). This is the first report of an antifreeze protein altering the T_m of a membrane, and implies a protein-lipid interaction.

The nature of that interaction is suggested by recent experiments. When the fluorescence anisotropy was done with DPH, a fluorescent probe that is located in the hydrophobic core of the bilayer, a significant increase in T_m of the lipid hydrocarbon chains was observed in the presence of the AFP (Fig. 8). By contrast, when the same study

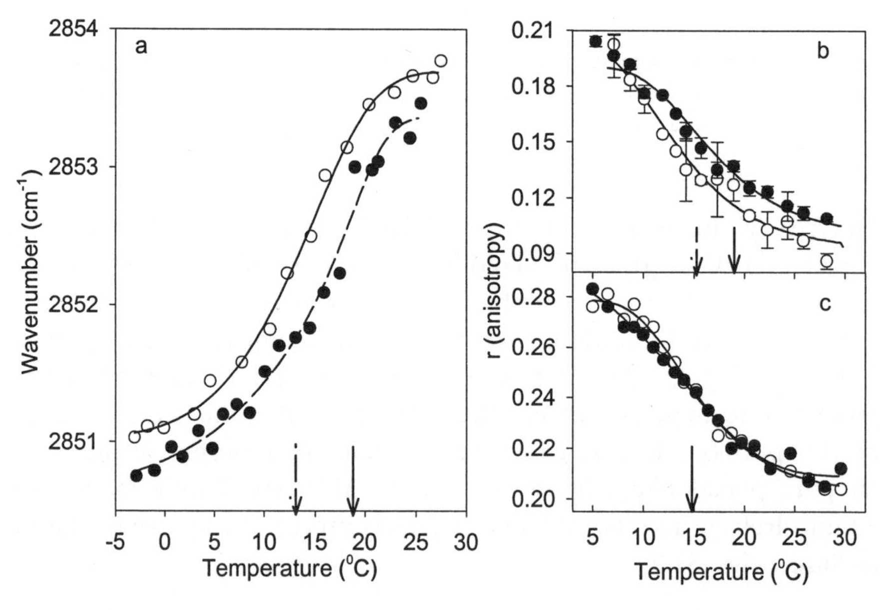

Fig. 8 AFP type I increases the phase transition of DGDG:DMPC liposomes. **(a)** FTIR determination of the phase transition of DGDG:DMPC liposomes in the presence (black circles) or absence (white circles) of AFP type I during cooling. T_m are indicated by arrows. **(b)** and **(c)** Anisotropy of DGDG:DMPC liposomes in the presence or absence of AFP type I as determined by diphenylhexatriene (b), and trimethylammonium diphenylhexatriene (c) as the liposomes are warmed through T_m. Symbols are the same as in panel **(a)**.

was repeated with TMA-DPH, a derivative of DPH that is located at the interfacial region of the bilayer, between the aqueous phase and hydrocarbon core, anisotropy in the presence of the protein was unchanged. This indicates that AFP type I, which is added peripherally to the liposomes, is affecting the order of the hydrophobic core of the bilayer, but not by interacting with the bilayer at the aqueous interface, as we had expected. Indeed, the anisotropy data are most consistent with insertion of the peptide into the bilayer, a finding that seems counter-intuitive.

We looked for further evidence that AFP type I altered the hydrophobic core of the bilayer by monitoring with FTIR the CH_2 scissoring vibration of the acyl chains, which describes the acyl chain packing order. We found that the AFP induces a significant alteration in the packing order (MMT *et al.*, submitted). Taken together, these data suggest a hydrophobic interaction between the peptide and the bilayer. This rather surprising result seems to be consistent with recent structural data on AFP type I suggesting that its interaction with ice is hydrophobic.[44–47]

One further piece of evidence confirms without much question that the AFPs interact directly with liposomes containing plant galactolipids. These experiments utilized a new liposome pelleting technique,[48] in which 100 nm liposomes can be pelleted with a 30 s, 50×g centrifugation and, therefore, any protein that may be weakly associated with the liposomes would remain associated after such a gentle centrifugation. Liposomes were incubated with AFP type I for an hour at room temperature or on ice, and then the pellet and supernatants were probed for the peptide. Approximately 25% of the peptide was in the pellet after the room temperature incubation, but 75% of the peptide was found in the pellet after an hour on ice. This shows that the peptide associates with the membrane after it has passed through T_m and that this interaction is stable, because the peptides maintain their membrane association after the samples are rewarmed prior to analysis (MMT *et al.*, submitted).

To our knowledge, there has been no investigation into whether AFPs affect the hydrophobic portion of cellular membranes or liposomes other than those described here. Therefore, we cannot state whether a shift in T_m and changes in the acyl chain packing order are general mechanisms of AFP mediated-membrane stabilization at low temperature. This study does, however, begin to give us insight into how the peptides may be interacting with membranes.

The mechanism of membrane stabilization by AFP type I appears to be different from the protection afforded to phospholipid liposomes by AFGPs, which we do not think insert into the bilayer. The optimal weight ratio of AFGP: lipid in leakage experiments[18] and solution NMR data[49] suggest that the AFGPs form a monolayer coverage of the liposomes to prevent leakage. This layer of peptides may block the transient leakage that would normally occur across a bilayer as it is cooled through its T_m. We present a schematic model for how the AFGPs and AFP type I may be interacting with the membranes to inhibit leakage in Fig. 9.

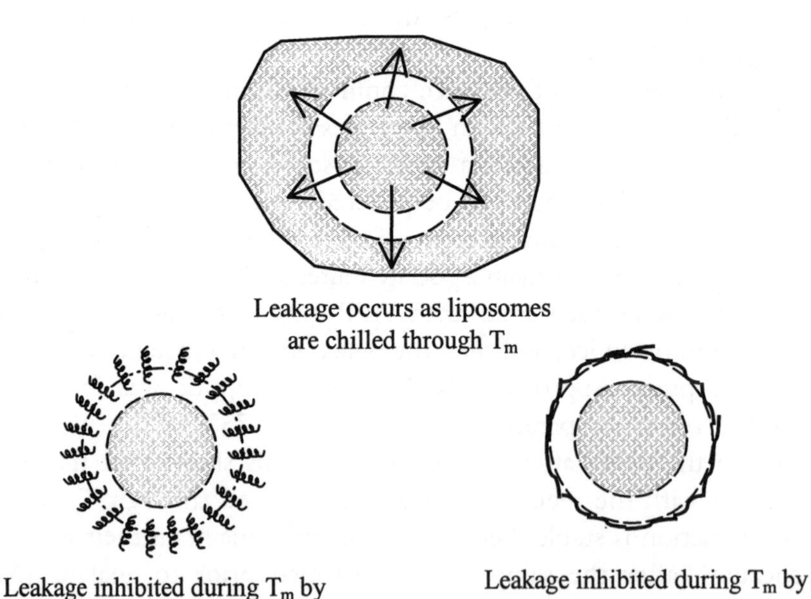
Leakage occurs as liposomes
are chilled through T_m
Leakage inhibited during T_m by
insertion of AFP type I into the bilayer
Leakage inhibited during T_m by
monolayer coverage of AFGPs

Fig. 9 Proposed mechanisms for the inhibition of leakage from membranes by AFP type I and AFGPs.

It is clear from the comparison of the chilling studies with pure phospholipid liposomes and galactolipid-containing liposomes that lipid composition plays an important role in determining whether the AFPs and AFGPs will protect, have no effect on, or damage membranes during chilling. It is important to use caution when comparing the results from one study to those of another, where a different membrane or lipid is employed, until we have a better understanding of the details regarding the mechanism(s) of interaction.

Do Antifreeze (Glyco)proteins Interact with Lipids and/or with Membrane Proteins?

It is unclear if the two proposed mechanisms of antifreeze protein-membrane protection are mutually exclusive. It is clear that a complete inhibition of leakage from membranes composed solely of lipids chilled through their phase transition can be achieved with AFGPs and AFP type I[18] (MMT *et al.*, submitted). This indicates that lipids are a main site of antifreeze protein action/interaction during low temperature stress. There may also be interaction between the antifreeze proteins and integral membrane proteins, which blocks leakage during chilling. However, there is no direct evidence for such an interaction.

Future Directions

Several groups have transformed freeze-sensitive plants with the gene for AFP types I and III in hopes of making a more freeze-tolerant or freeze-resistant plant.[50–53] These attempts have been met with little success. The results from Hincha[38] show that AFP types I and III damage thylakoids after freezing, so it is not surprising that the transformed plants are no more freeze-tolerant than the controls. However, AFP type I may improve the chilling tolerance of sensitive plants by stabilizing membranes as they pass through their phase transition. Achieving low temperature tolerance is a complicated process, and one protein certainly may not counter all of the deleterious effects of chilling. However, it could be a starting point for improving cold tolerance.

The studies on oocyte preservation[17,34] are important for the livestock breeding industry, where virtually all of the breeding is achieved by artificial insemination. Long-term, low temperature storage of livestock oocytes could ensure that genetically diverse and high quality animal lines are preserved. Preservation of oocytes also has obvious applications for human *in vitro* fertilization.

Organs for transplant might have a longer useful life if they could be preserved at low temperature with AF(G)Ps. So far, it appears that other additives will be needed to make cold storage of organs feasible.[35] Storing human platelets in blood banks at low temperature is also a possible use for AF(G)Ps, as discussed earlier. The results from Tablin *et al.* (1996)[8] show that platelets can be stored in the presence of AFGPs for 21 days at 4°C, and indicate that long-term, low temperature storage of human platelets is a real possibility.

The limitation to these applications is that the main commercial source of AFPs is the fish themselves, and currently the fish are the only source for AFGPs. Biological variation between fish and differences in isolation procedure can lead to significant differences in low temperature-membrane protection (MMT *et al.*, unpublished results; LM Hays, personal communication). Clearly, commercially viable applications await large scale production of the proteins, whether it be by chemical synthesis or by *in vitro* expression.

Acknowledgments

We are grateful to Drs. RE Feeney, LM Hays and DK Hincha for many insightful discussions on membrane stabilization by antifreeze proteins. We thank Drs. LM Crowe and LM Hays for providing figures to us.

References

1. Raison JK and Lyons JM (1986). Chilling Injury: A Plea for Uniform Terminology. *Plant Cell Environ.* 9: 685–686.

2. White JG and Krivit W (1967). An ultrastructural basis for the shape changes induced in platelets by chilling. *Blood* **30**: 625–635.
3. Zucker M and Borrelli J (1954). Reversible alterations in platelet morphology produced by anticoagulants and cold. *Blood* **28**: 602–608.
4. Raison JK and Orr (1986). Phase transitions in liposomes formed from the polar lipids of mitochondria from chilling-sensitive plants. *Plant Physiol.* **81**: 807–811.
5. Crowe JH, Hoekstra FA, Crowe LM, Anchordoguy TJ and Drobnis E. (1989). Lipid phase transitions measured in intact cells with Fourier transform infrared spectroscopy. *Cryobiology* **26**: 76–84.
6. Orr and Raison, JK (1990). The effect of changing the composition of phosphatidylglycerol from thylakoid polar lipids of oleander and cucumber on the temperature of the transition related to chilling injury. *Planta* **181**: 137–143.
7. Arav A, Zeron Y, Leslie SB, Behboodi E, Anderson GB and Crowe JH. (1996). Phase transition temperature and chilling sensitivity of bovine oocytes. *Cryobiology* **33**: 589–599.
8. Tablin F, Oliver AE, Walker NJ, Crowe LM and Crowe JH (1996). Membrane phase transition of intact human platelets: correlation with cold-induced activation. *J. Cell. Physiol.* **168**: 305–313.
9. Crowe JH, Tablin F, Tsvetkova NM, Oliver AE, Walker NJ and Crowe LM (1999). Are lipid phase transitions responsible for chilling damage in human platelets? *Cryobiology* **38**: 180–191.
10. Hazel JR (1995). Thermal adaptation in biological membranes: is homeovviscous adaptation the explanation? *Ann. Rev. Physiol.* **57**: 19–42.
11. Blok MC, van Deenen LLM and De Gier J. (1976). Effect of the gel to liquid crystalline phase on the osmotic behavior of phosphatidylcholine liposomes. *Biochim. Biophys. Acta* **433**: 1–12.
12. Papahadjopoulos D, Jacobson K, Nir S and Isac T (1973). Fluorescence polarization and permeability measurements concerning the effect of temperature and cholesterol. *Biochim. Biophys. Acta* **311**: 330–348.
13. Clerc SG and Thompson TE (1995). Permeability of dimyristoylphosphatidylcholine/dipalmitoylphosphatidylcholine bilayer membranes with coexisting gel and liquid-crystalline phases. *Biophys. J.* **68**: 2333–2341.
14. Quinn PJ (1985). A lipid-phase separation model of low-temperature damage to biological membranes. *Cryobiology* **22**: 128–146.

15. Chapman D (1975). Phase transitions and fluidity characteristics of lipids and cell membranes. *Quarterly Rev. Biophys.* **8**: 185–235.
16. Brown DA and London E (1998). Structure and origin of ordered lipid domains in biological membranes. *J. Mem. Biol.* **164**: 103–114.
17. Rubinsky B, Arav A, Mattioli M and DeVries AL (1990) The effect of antifreeze glycopeptides on membrane potential changes at hypothermic temperatures. *Biochem. Biophys. Res. Comm.* **173**: 1369–1374.
18. Hays LM, Feeney RE, Crowe LM, Crowe JH and Oliver AE (1996). Antifreeze glycoproteins inhibit leakage from liposomes during thermotropic phase transitions. *Proc. Natl. Acad. Sci. USA* **93**: 6835–6840.
19. Lewis RNAH and McElhaney RN (1991). The mesophoric phase behavior of lipid bilayers. In: Yeagle P (ed.) *The Structure of Biological Membranes.* CRC Press, Boca Raton, pp. 73–155.
20. Gennis RB (1989). The structures and properties of membrane lipids. In Gennis RB (ed.) *Springer Advanced Texts in Chemistry. Biomembranes: Molecular Structure and Function.* Springer-Verlag, New York, pp. 36–84.
21. Hazel JR and Williams EE (1990). The role of alterations in membrane lipid composition in enabling physiological adaptation of organisms to their physical environment. *Prog. Lipid Res.* **29**: 167–227.
22. Papahadjopoulos D, Vail WJ, Newton C, Nir S, Jacobsen K, Poste G, and Lazo R (1977). Studies on membrane fusion. III. The role of calcium-induced phase changes. *Biochim. Biophys. Acta* **465**: 579–598.
23. Mabrey S, Mateo PL and Sturtevant JM (1978). High-sensitivity scanning calorimetric study of mixtures of cholesterol with dimyristoyl- and diplamitoylphsophatidylcholines. *Biochemistry* **17**: 2464–2468.
24. Mantsch HH and McElhaney RN (1991). Phospholipid phase transitions in model and biological membranes as studied by infrared spectroscopy. *Chem. Phys. Lipids* **57**: 213–226.
25. Lentz BR (1993). Use of fluorescent probes to monitor molecular order and motions within liposome bilayers. *Chem. Phys. Lipids* **64**: 99–116.
26. Worrall D, Elias L, Ashford D, Smallwood M, Sidebottom C, Lillford P, Telford J, Holt C and Bowles D (1998). A carrot leucine-rich-repeat protein that inhibits ice recrystallization. *Science* **282**: 115–117.
27. Smallwood M, Worrall D, Byass L, Elias L, Ashford D, Doucet C J, Holt C, Telford J, Lillford P and Bowles DJ (1999). Isolation and characterization of a novel antifreeze protein from carrot (*Daucus carota*). *Biochem. J.* **340**: 385–391.

28. Antikainen M and Griffith M (1997). Antifreeze protein accumulation in freezing-tolerant cereals. *Physiol. Plant.* **99**: 423–432.
29. Duman J (1994). Purification and characterization of a thermal hysteresis protein from a plant, the bittersweet nightshade *Solanum dolcamara*. *Biochim. Biophys. Acta* **1206**: 129–135.
30. Urrutia ME, Duman JG and Knight CA (1992). Plant thermal hysteresis proteins. *Biochim. Biophys. Acta* **1121**: 199–206.
31. Tursman D and Duman JG (1995). Cryoprotective effects of thermal hysteresis protein on survivorship of frozen gut cells from the freeze-tolerant centipede *Lithobius forficatus*. *J. Expt. Zool.* **272**: 249–257.
32. Fahning ML and Garcia MA (1992). Status of cryopreservation of embryos from domestic animals. *Cryobiology* **29**: 1–18.
33. Davies PL and Sykes BD (1997). Antifreeze proteins. *Curr. Opin. Struct. Biol.* 7: 828–834.
34. Rubinsky B, Arav A and Fletcher GL (1991a). Hypothermic protection — A fundamental property of antifreeze proteins. *Biochem. Biophys. Res. Comm.* **180**: 566–571.
35. Lee CY, Rubinsky B and Fletcher GL (1992). Hypothermic preservation of whole mammalian organs with antifreeze proteins. *Cryo Lett.* **13**: 59–66.
36. Gong Z, Ewart KV, Hu Z, Fletcher GL and Hew CL (1996). Skin antifreeze protein genes of the winter flounder, *Pleuronectes americanus*, encode distinct and active polypeptides without the secretory signal and prosequences. *J. Biol. Chem.* **271**: 4106–4112.
37. Itskovitz-Eldor J, Levron J, Arav A, Bar-Ami S, Stein DW, Fletcher GL and Rubinsky B (1993). Hypothermic preservation of human oocytes with antifreeze proteins from sub-polar fish. *Cryo Lett.* **14**: 235–242.
38. Hincha DK, DeVries AL and Schmitt JM (1993). Cryotoxicity of antifreeze proteins and glycoproteins to spinach thylakoid membranes — comparison with cryotoxic sugar acids. *Biochim. Biophys. Acta* **1146**: 258–264.
39. Carpenter JF and Hansen TN (1992). Antifreeze protein modulates cell survival during cryopreservation: mediation through influence on ice crystal growth. *Proc. Natl. Acad. Sci. USA* **89**: 8953–8957.
40. Douce R and Joyard J (1990). Biochemistry and function of the plastid envelope. In: Palade GE (ed.) *Ann. Rev. Cell Biol.* **6**: 173–216.
41. Rubinsky B, Mattioli M, Arav A, Barboni B and Fletcher GL (1992a). Inhibition of Ca^{2+} and K^{+} currents by antifreeze proteins. *Am. J. Physiol.* **262**: R542–R545.

42. Negulescu PA, Rubinsky B, Fletcher GL and Machen TE (1992). Fish antifreeze proteins block Ca entry into rabbit parietal cells. *Am. J. Physiol.* **263**: C1310–C1313.
43. Oliver AE, Tablin F, Walker N and Crowe JH (1999). The internal calcium concentration of human platelets increases during chilling. *Biochim. Biophys. Acta* **1416**: 349–360.
44. Chao H, Houston ME, Hodges RS, Kay CM, Sykes BD, Loewen MC, Davies PL and Soennichsen FD (1997). A diminished role for hydrogen bonds in antifreeze protein binding to ice. *Biochemistry* **36**: 14652–14660.
45. Haymet ADJ, Ward LG, Harding MM and Knight CA (1998). Valine substituted winter flounder "antifreeze": preservation of ice growth hysteresis. *FEBS Lett.* **430**: 301–306.
46. Haymet ADJ, Ward LG and Harding MM (1999). Winter flounder "antifreeze" proteins: synthesis and ice growth inhibition of analogues that probe the relative importance of hydrophobic and hydrogen-bonding interactions. *J. Am. Chem. Soc.* **121**: 941–948.
47. Baardsnes J, Kondejewski LH, Hodges RS, Chao H, Kay C and Davies PL (1999). New ice-binding face for type I antifreeze protein. *FEBS Lett.* **463**: 87–91.
48. Zhan H (1999). A method for quick measurement of protein binding to unilamellar vesicles. *J. Biochem. Biophys. Meth.* **41**: 13–19.
49. Lane AN, Hays LM, Feeney RE, Crowe LM and Crowe JH (1998). Conformational and dynamic properties of a 14-residue antifreeze glycopeptide from Antarctic cod. *Protein Sci.* 7: 1555–1563.
50. Kenward KD, Altschuler M, Hildebrand D and Davies PL (1993). Accumulation of type I fish antifreeze protein in transgenic tobacco is cold-specific. *Plant Mol. Biol.* **23**: 377–385.
51. Kenward KD, Brandle J, McPherson J and Davies PL (1999). Type II fish antifreeze protein accumulation in transgenic tobacco does not confer frost resistance. *Transgenic Res.* **8**: 105–117.
52. Cutler AJ, Saleem M, Kendall E, Gusta LV, Georges F and Fletcher GL (1989). Winter flounder antifreeze protein improves the cold hardiness of plant tissues. *J. Plant Physiol.* **135**: 351–354.
53. Hightower R, Baden C, Penzes E, Lund P and Dunsmuir P (1991). Expression of antifreeze proteins in transgenic plants. *Plant Mol. Biol.* **17**: 1013–1022.

Chapter 9

Antifreeze Protein Gene Transfer in Salmonids

Woon-Kai Low
*Department of Biochemistry
University of Toronto
Toronto, Canada

Choy L Hew*
Department of Biological Sciences
National University of Singapore, Singapore

Margaret Shears and Garth Fletcher
Ocean Science Centre, Memorial University of Newfoundland
St John's, Newfoundland, Canada

Peter L Davies
Department of Biochemistry, Queen's University
Kingston, Ontario, Canada

Introduction

The generation of transgenic organisms involves the transfer of genes encoding a specific trait, or phenotype from one species or organism into another. In many instances, the technology is used either to boost the production level of an endogenous gene product, such as the growth hormone or other growth factors, by regulating its level or site of expression, or to incorporate a novel gene to confer a new and desirable characteristic. The intended purpose in commercial applications is usually to create an increase in efficiency of production or an improvement in product quality.

Transgenic technology, or the incorporation of foreign genes into an organism, requires careful selection of the candidate genes and their manipulation into gene constructs suitable for transfer, an appropriate mode of gene transfer and a fitting choice of host organism to achieve a practical end. Furthermore, other considerations such as the origin and safety of the candidate genes and the promoter sequences must also be considered, especially if the desired result is to be a viable commercial product.

A number of transgenic studies have involved the transfer of genes encoding antifreeze proteins from plants, insects (where they are known as thermal hysteresis proteins, THPs) or marine fish.[1–8] These proteins have a wide diversity in structure, but all act in a similar manner to prevent ice crystal growth.[9–14] The aim in most instances is to transfer the freeze or frost resistance capabilities from one organism into a more commercially valued organism that is susceptible to freezing stresses.

Rationale for AFP Gene Transfer to Atlantic Salmon

In many geographic locations, harsh environmental factors hamper or preclude a viable agricultural or aquacultural operation. One example is the low seawater temperatures in Atlantic Canada during the winter season which have made sea cage farming or sea-pen culture of several commercial fish species impractical. In the winter, near surface water temperatures can reach as low as –1.9°C and fish such as the Atlantic salmon (*Salmo salar*) which are not able to overcome temperatures below –0.7 to –0.8°C normally migrate to avoid such conditions. One approach to sea cage farming of Atlantic salmon has been to try and shorten the growth time of the fish, thus obviating the need for over winter culturing. The other approach is to create a freeze-tolerant salmon. The antifreeze protein, is obviously an excellent candidate for gene transfer studies for the latter approach.

Early studies demonstrated that the type I AFP was the sole "effector molecule" of freeze resistance in the winter flounder. This was accomplished by direct injection of AFP into rainbow trout that were acclimatized to seawater.[15] Rainbow trout are normally not freeze-resistant, but the trait was conferred to the injected fish, and resistance to −1.4°C was generated. Furthermore, the degree of freeze resistance was also dependent on the amount of AFP injected. These early experiments demonstrated that the introduction of AFPs, from one species of fish (winter flounder) into a second species (rainbow trout) was feasible. Based on these results, it was concluded that transfer of a single gene from the winter flounder could confer freeze resistance in other species.

For the past two decades, our laboratories have been actively involved in the transfer of the freeze resistance phenotype of the winter flounder (*Pleuronectes americanus*) to the Atlantic salmon. The effector molecule of the winter flounder's freeze resistance is a type I antifreeze polypeptide (AFP). This chapter will review the work performed to date on the transfer of a type I AFP gene from the winter flounder into the Atlantic salmon. The review will also attempt to present the project with a historical perspective, while data concerning the transfer of the AFP gene into Atlantic salmon will be described in a manner that reflects the challenges that were faced, and how they have been addressed in the past two decades.

Gene Transfer and Micro-Injection

The transfer of the AFP gene from winter flounder into Atlantic salmon began in 1982 with the injection of linearized AFP DNA into fertilized Atlantic salmon eggs. When these studies began, the transfer of foreign DNA into embryos was a relatively new technology, while the injection of foreign DNA into fish embryos posed some unique obstacles. The eggs of salmonids are surrounded by chorionic membranes that are opaque, and have no visibly distinct pronucleus. When the egg is fertilized in water, the chorion hardens, making penetration of the egg with a fine glass needle difficult. With the technology available at the time, the injection of foreign DNA into the pronucleus and subsequent incorporation into the host chromosome was a difficult task and new methodologies had to be developed. The method developed involved direct cytoplasmic injection of the foreign DNA into fertilized, but pre-activated salmon eggs. Foreign DNA was injected near the micropyle which is located in close proximity to the egg nucleus and can be easily visualized with a dissecting microscope.[16] Hence, the injection of foreign DNA near the micropyle delivers the DNA in close proximity to the developing pronuclei. After a few years were spent in developing this technique, the subsequent work was based on fish injected in 1985. Approximately 1×10^6 linearized DNA molecules were injected into 1800 salmon eggs that had a survival rate of approximately 80%. Analysis of the injected salmon showed a gene transfer success rate of approximately 3% with 1% of transgenic positive fish expressing AFP.[16–18]

The gene chosen for transfer was the winter flounder genomic clone, 2A-7,[19] (GenBank #M62315). The DNA used for injection consisted of the 2A-7 genomic clone linearized at the unique *Eco*RI restriction enzyme site, causing injected DNA to contain pUC19 vector DNA as well. This genomic clone contained the entire gene sequence for wflAFP-6, the major serum AFP of the winter flounder, as well as all *cis*-acting factors in transcription, even though several of these *cis*-acting factors had not yet been identified at that time. Without any prior knowledge of the mechanisms controlling expression, it was hoped that injection of the entire gene would allow for the production of functional AFP in Atlantic salmon, and lead to freeze-resistant fish.

More recently, work from our laboratories has demonstrated the presence of a sub-class of antifreeze polypeptides in winter flounder, known as skin-type AFPs.[20] These skin-type AFPs are more ubiquitously expressed than the liver-type AFPs, and have a much wider tissue distribution. Skin-type AFPs have also been found in two species of sculpins,[21,22] alluding towards a more prominent role in freeze protection. At the start of these transgenic studies, only the liver-type AFPs were known, and it was believed that production of AFPs in the liver was sufficient to provide freeze resistance.

Gene Integration and Inheritance

One of the key goals in creating a transgenic organism is to achieve stable integration of the transgene within the host genome into a host chromosome. Furthermore, the gene must be transferred to offspring in a Mendelian manner, so that future generations possess the desired characteristics. One of the difficulties encountered with foreign DNA injection is the lack of control over the integration site in the host genome. Thus, the effects of the DNA surrounding the integration site on expression of the transgene cannot be predicted. Furthermore, random integration of the transgene may have adverse effects on host genes as well. Thus, injected fish have to be manually screened to identify those individuals that contain the transgene, and subsequently subjected to further screening for stable transgene integration.

Early work using genomic Southern blotting techniques identified the presence of the 2A-7 gene in positive transgenic salmon.[17,18] Later screenings using PCR techniques identified several more positive fish.[22] The transgenic salmon were screened for expression of the transgene by immunoblotting serum samples with type I AFP-specific antibodies. Immunoblotting results showed that approximately 40% of the transgenic salmon identified by Southern/PCR screening produced AFPs in the serum.[18,22] Putative founder fish were then crossed with wild-type fish to produce both F1 and F2 offspring to examine the inheritance, and hence integration stability of the transgene. Inheritance in the F1 offspring varied from 17% to 64%, indicating that the original transgenic salmon were germ-line mosaic. However, crossing the F1 generation with wild-type produced close to 50% inheritance in the F2 generation, indicating stable integration and Mendelian inheritance of the transgene (Table 1).[18,22]

Further demonstration of stable integration and Mendelian inheritance has come recently from analysis of the F3 generation.[23,24] One male founder fish (#1469) that demonstrated the presence of the transgene was crossed with a wild-type female to generate a F1 generation that demonstrated only 17% inheritance. The F1 male, ASC96 was crossed again with a wild-type female to generate the F2 generation. The F2 generation demonstrated 50% inheritance, indicative of Mendelian genetics. Finally, the F2 male, ASC1313, was crossed with a wild-type female to generate F3 (#7000 series) with ~50% inheritance, and it was also back-crossed with an F1 female (ASC239) to generate

Table 1 Mendelian inheritance of AFP transgenes and their expression in Atlantic salmon.

Gene	Year	Generation	Cross	Transgene (% offspring)		Expression (% offspring)
				Observed	Expected	
	1985	P_1	Founder	3%	–	1%
	1989	F_1	$P_1 \times$ Wild	17%	–	17%
AFP	1990	F_2	$F_1 \times$ Wild	51–54%	50%	All transgenics
	1992	F_3	$F_2 \times$ Wild	52%	50%	All transgenics
	1992	F_3	$F_2 \times F_2$ (4)	69–77%	75%	All transgenics

an F3 generation (#8000 series) with 72% inheritance. These results provided conclusive evidence of the stable integration of the transgene into the salmon genome. Subsequent analysis of the F3 by Southern blotting analysis confirmed that only one copy of the transgene is present. With the heritability of the transgene established, studies were undertaken to identify the site of integration using linker-mediated nested PCR performed on fish from both #7000 and #8000 series F3 generation.[23] The sequences obtained from the two fish were identical, and identified 689 base pairs of unique sequence from the salmon genome. The identity of integration sites between the two F3 generation series also demonstrated the stable integration of the transgene.

Transgene Expression and Protein Activity

Beyond the transfer of a gene from one organism into another with stable integration, proper expression, processing and post-translational modification of the gene product to create a functional protein is also clearly required. The analysis of transgene expression in transgenic positive salmon has essentially been monitored by immunoblotting analysis for protein production. This section will focus on the results and conclusions that have been derived from assays for the expression of the transgene.

In winter flounder, wflAFP-6 (encoded by 2A-7) is expressed solely in the liver and is secreted into the serum.[20,25] The polypeptide is first secreted as a proAFP and consequently processed into the mature form in the serum by an unknown mechanism.[26–29] Early immunoblotting analysis demonstrated that not all fish that were positively transgenic, as identified by Southern blotting, were actually producing AFP at detectable levels in the serum.[30] It was also quite clear from the early immunoblotting analysis that even those fish that were producing the protein and properly secreting the AFP into the serum, were producing the protein in the proAFP form.[16,30,31] The Atlantic salmon clearly lacks the correct processing enzymes to create the mature wflAFP-6. It has been demonstrated that the proAFP is about 70% as active as the mature form.[28] On hindsight, lack of processing of the proAFP form may have been avoided through engineering of the gene to be transferred, but as mentioned previously, when these studies began, very little was known at the time about the actual expression of the wflAFP-6 gene and processing of the gene product.

In the winter flounder, wflAFP-6 levels in the serum are elevated in the winter season. The seasonal regulation is controlled by growth hormone, whose levels are related to day length.[26,32–38] Furthermore, expression of wflAFP-6 occurs only in the liver of the fish, resulting from transcriptional control residing in *cis*-acting factors within the single intron of the wflAFP-6 gene.[39–41] Northern blotting analysis of both the 7000 and 8000 series of the F3 generation showed that the wflAFP-6 corresponding mRNA could only be detected in liver tissue.[23] Additionally, when F3 fish were analyzed for AFP levels each month of the year, a significant increase was detected in November, with levels falling off by January.[23] These results may indicate that the regulatory elements of the wflAFP-6 gene are also functional in the Atlantic salmon. At the time of initiation of these studies, very little was known about the gene regulation of wflAFP-6 expression, and in this instance it was fortuitous that the entire gene was injected.

The winter flounder normally possesses approximately 40 copies of the liver-type AFP genes and approximately 40 copies of skin-type AFP genes. It is this high gene copy number, along with transcriptional control mechanisms and high mRNA stability,[42] that allow for a very high concentration of AFP in the serum during the winter season. Serum concentrations of AFPs reach a peak of approximately 10–15 mg/ml. Intuitively, it would seem that transgenic Atlantic salmon would also require high concentrations of AFPs in the serum to effect proper freeze resistance. However, as mentioned above, producing high concentrations of AFPs in salmon serum is already hampered by the fact that there may only be, at most, a few copies of the integrated transgene present. Furthermore, activity of the protein is also reduced by the lack of processing of the pro-sequence. Estimates for the amount of AFP found in transgenic salmon serum range from 0.07–20 μg/ml to 250 μg/ml,[18,23,30] clearly nowhere near the levels found in the winter flounder and around 100-fold lower than the minimal levels necessary for freeze resistance. However, the amount present is able to modify ice crystal morphology, as demonstrated by assays on F3 generation sera, albeit with no thermal hysteresis activity.[23]

Future Directions

Several hurdles remain in the creation of freeze-resistant Atlantic salmon that are related to issues which could not have been predicted when these studies were initiated. Some of these issues include the lack of processing of the pro-sequence, low gene copy number and hence low protein yield, and the lack of skin-type AFPs. Some of these problems could be addressed by further enhancing the current transgenic salmon. For example, the skin-type AFP genes could also be transferred into salmon. An alternative approach would be to generate entirely new fish harbouring engineered fish AFPs with higher activity levels, circumventing the need for increased protein yields. This approach may be feasible as antifreeze proteins do exist that are much more active than fish AFPs: for example those found in terrestrial insects, the THPs, which can provide freeze resistance to –4°C to –5°C.[13] However, introducing genes from insects into fish may create other difficulties such as an inability to properly fold and process the gene product, which would need to be considered.

When these transgenic studies were initiated, the field itself was a relatively new science. The generation of genetically modified organism is a fast growing field of research, and in these times it would be difficult to imagine the initiation of a transgenic project with as little information about the transgene as that available for the AFP gene when this project was initiated. While the generation of a fully freeze-resistant Atlantic salmon may eventually lead to the use of an entirely different gene, the transgenic fish generated in this study may yet be useful in other ways. Evidence exists that even low levels of AFPs can provide cold tolerance.[43] Furthermore, several lessons have been learned in the process over the past two decades that will be instrumental in the creation of future transgenic fish.

Prospects and Expectations for the Transgenics

The end of the 20th century has seen an increase in the number of attempts for the commercialization of new and emerging technologies. There is no doubt that transgenic technology can be used to produce superior fish for domestic consumption. The creation of faster growing fish[44–53] and the freeze-tolerant salmon represent two of the more well known examples of this technology. Such developments could impart significant benefits to the growing population[54,55] and, at the same time, have a positive impact on the stability and preservation of the marine environment. Nonetheless, there are several issues that deal with food safety, environmental impact and animal health that will need to be addressed.

Food safety is important in view of recent consumer concerns about genetically modified foods. However, it is the properties of the individual food rather than the process itself that need to be examined.[57] In many countries, there are government agencies that deal with these issues. For example, in Canada, the appropriate regulatory agency is Health Canada. The Novel Food regulations, part of the Canadian Food and Drug Act makes it mandatory to notify and obtain Health Canada approval before any novel food is sold or advertised for sale in Canada. The assessment will include the process used to develop the modified organism, the characteristic of the organisms, the nutritional quality, possibility of toxicants and the changes in allergenicity. In United States, the appropriate government agency is the Center for Veterinary Medicine within the Food and Drug Administration.

Presently, salmon are cultured in sea pens that are located in coastal waters. There is concern that the accidental escape of these transgenic fish might lead to interbreeding with the wild stocks. It is generally agreed that the table fish, if they are to be cultured in cages, must be sterile while the broodstock be maintained in secure, contained land-based facilities. The production of all-female triploids is presently one of the approaches.[56–59] Aquaculture development on land using closed circulating system has the additional advantages in disease control, feeding regiments, waste treatment and other manipulations.

The transgenic fish must be healthy and exhibit normal feeding and other behavior patterns typical of domesticated species. Lastly, the general public must be informed and educated that new products are safe, beneficial to their well-being and to the environment.

References

1. Rancourt DE, Walker VK and Davies PL (1987). Flounder antifreeze protein synthesis under heat shock control in transgenic *Drosophila melanogaster*. *Mol. Cell. Biol.* 7(6): 2188–2195.
2. Rancourt DE, Peters ID, Walker VK and Davies PL (1990). Wolffish antifreeze protein from transgenic Drosophila. *Biotechnology (NY)* **8**(5): 453–457.
3. Hightower R, Baden C, Penzes E, Lund P and Dunsmuir P (1991). Expression of antifreeze proteins in transgenic plants. *Plant Mol. Biol.* 17(5): 1013–1022.
4. Kenward KD, Altschuler M, Hildebrand D and Davies PL (1993). Accumulation of type I fish antifreeze protein in transgenic tobacco is cold-specific. *Plant Mol Biol* **23**(2): 377–385.

5. Peters ID, Rancourt DE, Davies PL and Walker VK (1993). Isolation and characterization of an antifreeze protein precursor from transgenic Drosophila: evidence for partial processing. *Biochimica et Biophysica Acta* **1171**(3): 247–254.
6. Walker VK, Rancourt DE and Duncker BP (1995). The transfer of fish antifreeze genes to Drosophila: a model for the generation of transgenic beneficial insects. *Proc. Entomol. Soc. Ontario* **126**: 3–13.
7. Wallis JG, Wang H and Guerra DJ (1997). Expression of a synthetic antifreeze protein in potato reduces electrolyte release at freezing temperatures. *Plant Mol. Biol.* **35**(3): 323–330.
8. Kenward KD, Brandle J, Macpherson J and Davies PL (April 1999). Type II fish antifreeze protein accumulation in transgenic tobacco does not confer frost resistance. *Transgenic Res.* **8**(2): 105–117.
9. Hew CL and Yang DS (1992). Protein interaction with ice. *Eur. J. Biochem.* **203**(1–2): 33–42.
10. Yeh Y and Feeney RE (1996). Antifreeze proteins: structures and mechanisms of function. *Chem Rev* **96**(2): 601–618.
11. Ewart KV, Lin Q and Hew CL (1999). Structure, function and evolution of antifreeze proteins. *Cell Mol. Life Sci.* **55**(2): 271–283.
12. Barrett J (2001). Thermal hysteresis proteins. *Int. J. Biochem. Cell Biol.* **33**(2): 105–117.
13. Duman JG (2001). Antifreeze and ice nucleator proteins in terrestrial arthropods. *Ann. Rev. Physiol.* **63**: 327–357.
14. Fletcher GL, Hew CL and Davies PL (2001). Antifreeze proteins of teleost fishes. *Ann. Rev. Physiol.* **63**: 359–390.
15. Fletcher GL, Kao MH and Fourney RM (1986). Antifreeze peptides confer freezing resistance to fish. *Can. J. Zool.* **64**(9): 1897–1901.
16. Shears MA, Fletcher GL, Hew CL, Gauthier S and Davies PL (1991). Transfer, expression, and stable inheritance of antifreeze protein genes in Atlantic salmon (Salmo salar). *Mol. Mar. Biol. Biotech.* **1**(1): 58–63.
17. Fletcher GL, Shears MA, King MJ, Davies PL and Hew CL (1988). Evidence for antifreeze protein gene transfer in Atlantic salmon (*Salmo salar*). *Can. J. Fish. Aquat. Sci.* **45**(2): 352–357.
18. Hew CL, Fletcher GL and Davies PL (1995). Transgenic salmon: tailoring the genome for food production. *J. Fish Biol.* **47A**: 1–19.
19. Davies PL and Gauthier SY (1992). Antifreeze protein pseudogenes. *Gene* **112**(2): 171–178.

20. Gong Z, Ewart KV, Hu Z, Fletcher GL and Hew CL (1996). Skin antifreeze protein genes of the winter flounder, *Pleuronectes americanus*, encode distinct and active polypeptides without the secretory signal and prosequences. *J. Biol. Chem.* **271**(8): 4106–4112.
21. Low WK, Miao M, Ewart KV, Yang DSC, Fletcher GL and Hew CL (1998). Skin-type antifreeze protein from the shorthorn sculpin, *Myoxocephalus scorpius. J. Biol. Chem.* **273**(36): 23098–23103.
22. Low WK, Lin Q, Stathakis C, Miao M, Fletcher GL and Hew CL (2001). Isolation and characterization of skin-type, type I antifreeze polypeptides from the longhorn sculpin, *Myoxocephalus octodecemspinosus. J. Biol. Chem.* **2**: 2.
23. Hew CL, Poon R, Xiong F, Gauthier S, Shears M, King M, Davies P and Fletcher G (1999). Liver-specific and seasonal expression of transgenic Atlantic salmon harboring the winter flounder antifreeze protein gene. *Transgenic Res.* **8**(6): 405–414.
24. Fletcher GL, Goddard SV and Wu Y (1999). Antifreeze proteins and their genes: from basic research to business opportunity. *Chemtech.* **29**(6): 17–28.
25. Gong Z, King MJ, Fletcher GL and Hew CL (1995). The antifreeze protein genes of the winter flounder, *Pleuronectus americanus*, are differentially regulated in liver and non-liver tissues. *Biochem. Biophys. Res. Commun.* **206**(1): 387–392.
26. Hew CL, Liunardo N and Fletcher GL (1978). *In vivo* biosynthesis of the antifreeze protein in the winter flounder — evidence for a larger precursor. *Biochem. Biophys. Res. Commun.* **85**(1): 421–427.
27. Davies PL, Roach AH and Hew CL (1982). DNA sequence coding for an antifreeze protein precursor from winter flounder. *Proc. Natl. Acad. Sci. USA* **79**(2): 335–339.
28. Pickett M, Scott G, Davies P, Wang N, Joshi S and Hew C (1984). Sequence of an antifreeze protein precursor. *Eur. J. Biochem.* **143**(1): 35–38.
29. Hew CL, Wang NC, Yan S, Cai H, Sclater A and Fletcher GL (1986). Biosynthesis of antifreeze polypeptides in the winter flounder. Characterization and seasonal occurrence of precursor polypeptides. *Eur. J. Biochem.* **160**(2): 267–272.
30. Hew CL, Davies PL and Fletcher G (1992). Antifreeze protein gene transfer in Atlantic salmon. *Mol. Mar. Biol. Biotech.* **1**(4–5): 309–317.

31. Hew CL, Du S, Gong Z, Fletcher GL, Shears M and Davies PL (1991). Biotechnology in aquatic sciences: improved freezing tolerance and enhanced growth in Atlantic salmon by gene transfer. *Bull. Inst. Zool. Academia Sinica* **Monograph 16:** 341–356.
32. Fletcher GL, Campbell CM and Hew CL (1978). The effects of hypophy-sectomy on seasonal changes in plasma freezing-point depression, protein "antifreeze," and Na^+ and Cl^- concentrations of winter flounder (*Pseudopleuronectes americanus*). *Can. J. Zool.* **56**(1): 109–113.
33. Hew CL and Fletcher GL (1979). The role of pituitary in regulating antifreeze protein synthesis in the winter flounder. *FEBS Lett.* **99**(2): 337–339.
34. Fletcher GL (1981). Effects of temperature and photoperiod on the plasma freezing point depression, chloride ion concentration and protein antifreeze in winter flounder (*Pseudopleuronectes americanus*)." *Can. J. Zool.* **59**(2): 193–201.
35. Pickett MH, Hew CL and Davies PL (1983). Seasonal variation in the level of antifreeze protein mRNA from the winter flounder. *Biochim Biophys Acta* **739**(1): 97–104.
36. Fourney RM, Fletcher GL and Hew CL (1984a). Accumulation of winter flounder antifreeze messenger RNA after hypophysectomy. *Gen. Comp. Endocrinol.* **54**(3): 392–401.
37. Fourney RM, Fletcher GL and Hew CL (1984b). The effects of long day length on liver antifreeze messenger RNA in the winter flounder, *Pseudopleuronectes americanus. Can. J. Zool.* **62**(8): 1456–1460.
38. Idler DR, Fletcher GL, Belkhode S, King MJ and Hwang SJ (1989). Regulation of antifreeze protein production in winter flounder: a unique function for growth hormone. *Gen Compar. Endocrinol.* **74**(3): 327–334.
39. Chan SL, Miao M, Fletcher GL and Hew CL (1997). The role of CCAAT/ enhancer-binding protein alpha and a protein that binds to the activator-protein-1 site in the regulation of liver-specific expression of the winter flounder antifreeze protein gene. *Eur. J. Biochem.* **247**(1): 44–51.
40. Miao M, Chan SL, Hew CL and Fletcher GL (1998a). Identification of nuclear proteins interacting with the liver-specific enhancer B element of the antifreeze protein gene in winter flounder. *Mol. Mar. Biol. Biotech.* 7(3): 197–203.
41. Miao M, Chan SL, Hew CL and Gong Z (1998b). The skin-type antifreeze protein gene intron of the winter flounder is a ubiquitous enhancer lacking a functional C/EBPα binding motif. *FEBS Lett.* **426**(1): 121–125.

42. Duncker BP, Koops MD, Walker VK and Davies PL (1995). Low temperature persistence of type I antifreeze protein is mediated by cold-specific mRNA stability. *FEBS Lett.* **377**(2): 185–188.
43. Wang R, Zhang P, Gong Z and Hew CL (1995). Expression of the antifreeze protein gene in transgenic goldfish (*Carassius auratus*) and its implication in cold adaptation. *Mol. Mar. Biol. Biotech.* **4**(1): 20–26.
44. Ozato K, Wakamatsu Y and Inoue K (1992). Medaka as a model of transgenic fish. *Mol. Mar. Biol. Biotechnol.* **1**(4–5): 346–354.
45. Gong Z, Du SJ, Fletcher G, Shears M, Davies P, Derlin B and Hew CL (1994). Biotechnology for aquaculture: Transgenic salmon with enhanced growth and freeze-resistance. *Proceedings of the International Symposium on Biotechnology Applications in Aquaculture, 10 December 1994.*
46. Devlin RH, Yesaki TY, Donaldson EM, Du SJ and Hew CL (1995). Production of germline transgenic Pacific salmonids with dramatically increased growth performance. *Can. J. Fish. Aquat. Sci. J. Can. Sci. Halieut. Aquat.* **52**(7): 1376–1384.
47. Martinez R, Estrada MP, Berlanga J, Guillen I, Hernandez O, Cabrera E, Pimentel R, Morales R, Herrera F, Morales A, Pina JC, Abad Z, Sanchez V, Melamed P, Lleonart R and de la Fuente J (1996). Growth enhancement in transgenic tilapia by ectopic expression of tilapia growth hormone. *Mol. Mar. Biol. Biotechnol.* **5**(1): 62–70.
48. Saunders RL, Fletcher GL and Hew CL (1998). Smolt development in growth hormone transgenic Atlantic salmon. *Aquaculture* **168**: 1–4.
49. Krasnov A, Agren JJ, Pitaknen TI and Molsa H (1999). Transfer of growth hormone (GH) transgenes into Arctic charr. (*Salvelinus alpinus* L.) II. Nutrient partitioning in rapidly growing fish. *Genet. Anal.* **15**(3/5): 99–105.
50. Mori T and Devlin RH (1999). Transgene and host growth hormone gene expression in pituitary and nonpituitary tissues of normal and growth hormone transgenic salmon. *Mol. Cell. Endocrinol.* **149**(1/2): 129–139.
51. Pitkanen TI, Krasnov A, Teerijoki H and Molsa H (1999). Transfer of growth hormone (GH) transgenes into Arctic char. (*Salvelinus alpinus* L.) I. Growth response to various GH constructs. *Genet. Anal.* **15**(3/5): 91–98.
52. Devlin RH, Swanson P, Clarke WC, Plisetskaya E, Dickhoff W, Moriyama S, Yesaki TY and Hew CL (2000). Seawater adaptability and

hormone levels in growth-enhanced transgenic coho salmon, *Oncorhynchus kisutch*. *Aquaculture* **191**(4): 367–385.

53. Martinez R, Juncal J, Zaldivar C, Arenal A, Guillen I, Morera V, Carillo O, Estrada M, Morales A and Estrada MP (2000). Growth efficiency in transgenic tilapia (*Oreochromis* sp.) carrying a single copy of an homologous cDNA growth hormone. *Biochem. Biophys. Res. Commun.* **267**(1): 466–472.
54. Serageldin I (1999). Biotechnology and food security in the 21st century. *Science* **285**(5426): 387–389.
55. Trewavas A (1999). Much food, many problems. *Nature* **402**(6759): 231–232.
56. Johnstone R (1996). Experience with salmonid sex reversal and triploidisation technologies in the United Kingdom. *Bull. Aquacul. Assoc. Canada* **96**(2): 9–13.
57. Devlin RH and Donaldson EM (1992). Containment of genetically altered fish with emphasis on salmonids. In: Hew CL and Fletcher GL (eds.) *Transgenic Fish*. World Scientific, pp. 229–265.
58. Benfrey TJ (1999). The physiology and behaviour of triploid fishes. *Rev. Fish. Sci.* **7**: 39–67.
59. Cotter D, O'Donovan V, O'Maoiléidigh N, Rogan G, Roche N and Wilkins N (2000). An evaluation of the use of triploid Atlantic salmon (*Salmo salar* L.) in minimising the impact of escaped farmed salmon on wild populations. *Aquaculture* **186**: 61–75.

Subject Index

Species Index

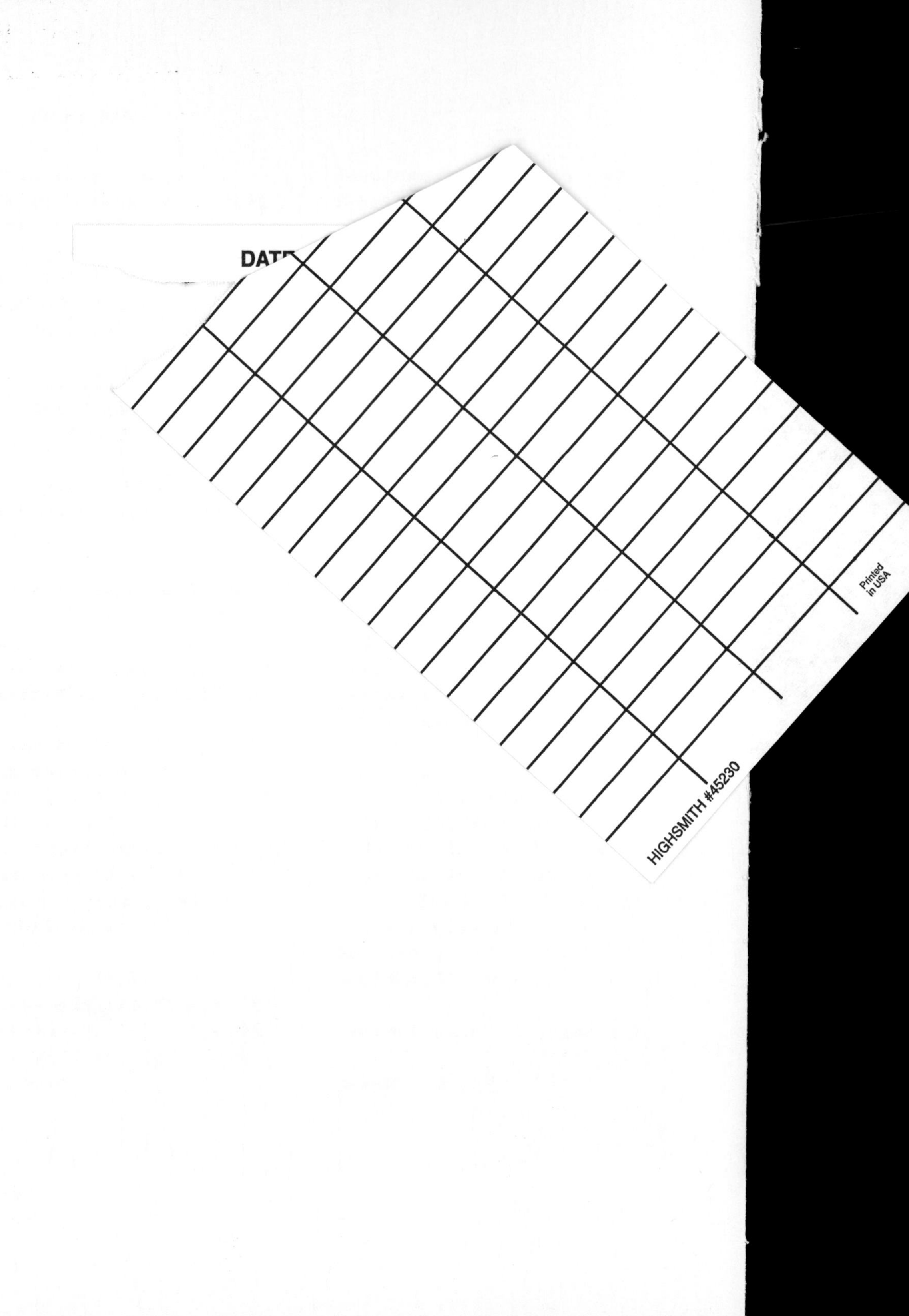

DAT
Printed in USA
HIGHSMITH #45230